吴平 ◎ 著

共建美丽中国
新时代生态文明理念、政策与实践

Building a Beautiful China Together:
Concepts, Policies & Practices on Ecological Civilization in the New Era

商务印书馆
The Commercial Press

图书在版编目(CIP)数据

共建美丽中国:新时代生态文明理念、政策与实践:汉、英/吴平著.—北京:商务印书馆,2018
ISBN 978-7-100-16508-2

Ⅰ.①共… Ⅱ.①吴… Ⅲ.①生态环境建设—研究—中国—汉、英 Ⅳ.①X321.2

中国版本图书馆 CIP 数据核字(2018)第 191229 号

权利保留,侵权必究。

共建美丽中国:新时代生态文明理念、政策与实践
吴平 著

商 务 印 书 馆 出 版
(北京王府井大街36号 邮政编码100710)
商 务 印 书 馆 发 行
北京顶佳世纪印刷有限公司印刷
ISBN 978-7-100-16508-2

2018年10月第1版　　　　　开本710×1000 1/16
2018年10月北京第1次印刷　印张20¾
定价:65.00元

目 录
CONTENTS

第一章　全面推进生态环境领域国家治理体系和治理能力现代化……………… 1
Chapter 1 Comprehensively Promote the Modernization of State Governance System and Governance Capabilities in the Field of Ecology & Environment ……………… 6

第二章　生态治理现代化的思想基础和理论共识……………………………… 13
Chapter 2 The Ideological Basis and Theoretical Consensus on Ecological Governance Modernization……………………………………………………………………… 17

第三章　生态治理需要强化法治思维…………………………………………… 22
Chapter 3 Strengthen the Rule of Law in Ecological Governance ………………… 26

第四章　生态治理体系的价值取向和立法路径………………………………… 32
Chapter 4 Value Orientation and Legislative Path of Ecological Governance System ……… 37

第五章　创新执法机制　护航绿色发展………………………………………… 43
Chapter 5 Innovate the Law Enforcement Mechanisms to Safeguard Green Development ………… 48

第六章　第三方评估促进生态治理现代化……………………………………… 56
Chapter 6 Utilize Third-party Evaluation to Promote Ecological Governance Modernization ……… 60

第七章　构建多元协同的生态治理模式………………………………………… 66
Chapter 7 Establish a Model of Ecological Governance: Coordinating All Stakeholders ……… 70

第八章　地方政府竞争转向助推生态治理……………………………………… 76
Chapter 8 Transform from Local Government Competition to Ecological Governance ………… 80

第九章　以生态红线为基准谋划国土空间开发………………………………… 85
Chapter 9 Develop Geographical Space on the Basis of Ecological Red Line ……………… 90

第十章　生态文明建设要算好自然资源这本账·········· 95
Chapter 10 Building an Ecological Civilization Entails a Clear Understanding of Natural Resources ·········· 100

第十一章　践行生态补偿　谋划绿富双赢·········· 107
Chapter 11 Win-win: Ecological Compensation for Wealth and Green Development ·········· 112

第十二章　技术创新引领绿色发展新动力·········· 119
Chapter 12 Technological Innovation: A New Driving Force for Green Development ·········· 124

第十三章　智能技术助力生态治理现代化·········· 131
Chapter 13 Smart Technologies Promote Modernization of Ecological Governance ·········· 135

第十四章　全球气候治理的经验与启示·········· 141
Chapter 14 Experiences and Inspiration from Global Climate Governance ·········· 145

第十五章　谱写绿色丝绸之路新篇章·········· 151
Chapter 15 A New Chapter for Green Silk Road ·········· 155

第十六章　国家公园：奏响生态治理现代化新乐章·········· 161
Chapter 16 National Parks Play an Important Role in Ecological Governance ·········· 166

第十七章　草原生态治理实现增收增绿·········· 172
Chapter 17 Grassland Ecological Governance Increases Income and Green Space ·········· 177

第十八章　恢复湿地以养护"地球之肾"·········· 185
Chapter 18 Restore Wetland to Protect "Kidney of the Earth" ·········· 189

第十九章　抚育经营森林　换得山绿民富·········· 195
Chapter 19 Nurture and Manage Forest to Make Mountains Green and People Rich ·········· 199

第二十章　创新沙漠治理　培育金色产业·········· 206
Chapter 20 Innovate Desert Governance and Cultivate Golden Industry ·········· 209

第二十一章　推进海洋生态治理　守护"蓝色家园"·········· 214
Chapter 21 Promote Marine Environmental Governance and Protect the "Blue Home" ·········· 219

第二十二章　治理雾霾须关注舆论的"风" …………………………………… 226
Chapter 22 Pay Attention to Public Opinions on Smog Control ………… 231

第二十三章　提升水治理能力　迎来河清海晏 ……………………………… 237
Chapter 23 Enhance Water Governance Capabilities ………………………… 241

第二十四章　防治土壤污染　守护沃土良田 ………………………………… 248
Chapter 24 Prevent and Control Soil Pollution to Safeguard Fertile Soil and Farmland ……… 253

第二十五章　攻坚面源污染　共建美丽乡村 ………………………………… 260
Chapter 25 Tackle Non-point Source Pollution to Build a Beautiful Rural China …… 265

第二十六章　打造特色小镇要坚持生态优先 ………………………………… 271
Chapter 26 Prioritize Ecology in Building Towns with Distinctive Features ………… 276

第二十七章　呼唤动物福利时代 ……………………………………………… 283
Chapter 27 The Era of Animal Welfare ………………………………………… 288

第二十八章　维护生态安全　推进永续发展 ………………………………… 294
Chapter 28 Maintain Ecological Security and Promote Sustainable Development ……… 299

第二十九章　电商进村入户念活绿色致富经 ………………………………… 306
Chapter 29 E-commerce Enters Villages to Foster Green Development …… 311

第三十章　环保督察整改倒逼绿色转型 ……………………………………… 317
Chapter 30 Environmental Inspection Drives Green Transformation ……… 321

第一章　全面推进生态环境领域国家治理体系和治理能力现代化

中共十八大将生态文明建设纳入中国特色社会主义事业"五位一体"总体布局，十八届三中全会将推进国家治理体系和治理能力现代化作为全面深化改革的总目标，这意味着推进生态环境领域国家治理体系和治理能力现代化有两重战略意义：一是有助于改善生态环境状况，促进经济社会系统与生态系统协同发展，实现人与自然和谐共存；二是有利于推进生态文明建设领域的国家治理体系和治理能力现代化。

"立志欲坚不欲锐，成功在久不在速。"生态文明建设对我国来说，既是机遇，又是挑战，要把握好这个机遇，赢得挑战，就需要采取一系列卓有成效的方法，而推进生态环境治理是生态文明建设的必由之路。当前，亟待厘清生态治理现代化的内涵、外延及其实现路径。

国家治理转型背景下的生态环境治理现代化

在当代中国的语境下，国家治理体系是指作为执政党的中国共产党带领全国人民有效治国理政的制度体系，国家治理能力是指用制度有效管理国家事务的能力。国家治理体系与治理能力的现代化，就是国家制度体系能够与时俱进地发展、转型，以支撑和适应中国的社会主义现代化事业。改革开放以来，我国实现了持续高速的经济增长，自2010年以来成为世界第二大经济体，国家治理的经济绩效得到全国人民的认可，也得到世界的关注。但是，GDP迅猛增长的代价却是生态环境质量恶化，并影响了人民群众的生活质量甚至生命健康。尽管自20世纪90年代以来保护生态的呼声越来越高，但生态环境的治理速度赶不上污染和环境损坏的速度，生态危机反而越来越严重，甚至对人们的健康生活乃至生存构成了威胁。中国不愿意也不能走"先污染后治理"的老路,"在发展中保护,在保护中发展"成了这个时代最响亮的主题之一。能否控制环境污染、建设生态文明,成为检验国家治理体系有效性的一个重要标志。

国家治理体系和治理能力现代化的提出,就是要加快制度建设的步伐,提高国家治理能力,以确保中国的现代化事业顺利推进,包括能够有效控制环境污染、实施生态保护、实现生态文明。如果污染失控、生态环境遭到破坏,这就表明国家治理存在不足,即国家治理出现了体制性困境,或者说国家治理体制存在严重缺陷。从宏观上看,中国在实现现代化进程中面临历史性的转型挑战,即现代化进程中伴随经济和社会关系发生重大结构性变迁而产生大量经济和社会的矛盾和冲突,但这些矛盾和冲突不能在经济领域和社会领域自我矫正,需要国家通过各种治理手段进行干预,因此,国家治理体系和治理能力现代化就是为了解决这些矛盾和冲突而做出的体制性调整和完善。

毫无疑问,生态治理现代化是国家治理现代化的重要内容,它的内涵可以从现实压力和对传统的反思两个方面来理解。从现实压力来看,生态危机已经成了当前我们面临的重要威胁,从过去的"民以食为天"变成了今天的"民以天为食"。近年来,各地生态环境事件不断发生,并牵涉政治、经济、社会等方面,生态问题甚至成为社会问题的导火索,如果没有好的生态,一切等于白费。从对传统的反思来看,生态危机的出现是传统发展方式的结果,要从根本上改变这种现状,就要改变发展方式,重新认识生态保护和经济发展的关系。

作为发展中国家,如果说在发展的早期,即便认识到了保护生态的重要性,但缺乏动力和能力去实践,因而处在"没有发展就无法保护,保护了就不能发展"的状态的话,那么现在就已经到了"没有保护就没有发展,不保护就等于发展为零,甚至发展为负"的地步。因此,现实的危机和对传统发展方式的反思推动着生态治理必须走向现代化。生态危机是转型危机的重要组成部分,相应地,生态治理现代化也是国家治理体系和治理能力现代化的重要组成部分。生态治理现代化一方面要保证治理行为本身的现代化——法治化、制度化、规范化、程序化、多元化，更为重要的是，生态治理现代化必然意味着生态治理能力的现代化。总之，生态治理必须要有现实性效果，必须要解决生态危机，必须要实现生态保护和经济发展的平衡，必须要确保中华民族的永续发展，必须要为世界生态问题的解决贡献我们独特的力量。而要做到这些，就必须从发展理念、法律制度、多元主体、技术支撑等方面做出改变，这个变革的过程，就是推进生态治理体系和治理能力的现代化的过程。

生态环境治理体系和治理能力现代化的实现路径

树立绿色发展理念，以绿色创新为发展动力，"绿水青山就是金山银山"。生态治理秉承现代化的生态观，要求转变发展动力，改变传统的依靠资源、资本、劳动扩张来发展经济的外生发展模式，以创新作为新的经济增长引擎，尤其是发挥绿色创新对经济增长的带动作用。绿色创新需要从理念和路径等多方位全面着手，树立创新的绿色理念，坚持"把良好生态环境作为公共产品向全民提供，努力建设一个生态文明的现代化中国"。绿色创新可以通过开发绿色产品，引进和创造绿色技术，开拓绿色消费市场，采用绿色资源，升级绿色管理制度、构建完善绿色金融体系等方式，提高经济运行的质量和效率，培育新的经济增长点，促进经济发展。

构建以法律为基础的生态环境治理体系。"法律是治国之重器，良法是善治之前提"，完善的法律体系可以明晰生态治理主体的权责，提高生态治理的认可度，保障生态治理体系的运行。因此，需要以生态环境产权制度为切入点，健全和完善生态环境保护、自然资源保护、污染防治、能源安全等重要领域的立法，完善循环经济、可再生能源等领域的生态协同性法律。同时，还要完善监督管理体制和问责机制，引进企业、民众、社会组织参与生态治理，形成全方位、多维度的生态治理制度体系。

不断提高生态治理能力。生态治理现代化，关键在于生态治理能力的现代化。提高生态治理能力水平，一要加强党对生态文明建设的领导，充分发挥政府集中力量办大事的优势。切实做到党政同责，一岗双责，齐抓共管，失职追责，并在治理过程中引入监督机制，保证治理行为本身符合相关法律规范。二要将地方政府竞争引导至有利于生态文明建设的方向，通过提高地方政府治理环境的收益支出比，引入生态政绩考核，执行地区内统一的环保标准，把生态环境优势转化为各地经济发展的真实竞争力，使得生态治理成为地方政府的常规性工作，从应急转向常态，切实保障治理本身的不断进行。三要不断加强绿色发展执法保障能力。一系列的生态法律，只有得到切实的执行，才能发挥有效作用，这不仅意味着要按照相关法律办事，而且意味着一旦出现违法行为，必须受到法律的制裁，在认识得到统一，法律制定完善之后，执行力就决定着治理的实际效果，因此，加强生态执法对于生态治理现代化具有不可替代的特殊意义。四要建立综合完善的生态环境评估指标，对治理结果进行第三方评估。从治理结果来看，生态治理能力的提高必须体现为生态环境的改善，这也是对治理过程的最终检验。因此，在生态治理中要坚持结果导向思维，发挥智库第三方评估的作用。第三方评估在生态治理中具有独特的作用，他们的参与是推动生态治理过程不断完善，生态治理能力不断提高的重要保证。

促进生态治理实践与科学技术的深度融合。"工欲善其事，必先利其器"，技术就是改造自然之"器"。生态治理现代化离不开科学技术的支撑，推进生态治理领域的科学研究，利用被证实的研究成果指导治理实践，坚持一切治理从实际出发，就能防止盲目治理，达到事半功倍的效果。因此，需要重视科学技术与生态治理融合，提升改造自然，进而与自然协调发展的能力。能源和通信技术的革新与结合，在人类经济转型过程中发挥着决定性作用。当前，我们正处在可再生能源与互联网信息技术相融合的新时代，利用大数据、云计算、"互联网+"等智能技术，大力发展风能、太阳能、生物质能等可再生能源，提高全社会能源系统的效率、绿色化水平，这不仅对生态治理现代化具有重要意义，从长远来看，也是推动经济转型升级的重要力量。

推进多元化主体协同共治。生态治理中的主体包括国家、企业、民众、各类社会组织等，其中国家作为主导，制定治理的政策，保证政策的贯彻落实；同时也要引导企业、民众、社会组织参与生态治理的决策，对治理过程进行监督，对治理绩效进行客观评价。发达国家现代化的成功经验和历史教训表明，仅仅依靠政府进行生态治理是不可能取得成功的。多元化主体参与式治理已经逐渐成为社会各界的广泛共识。鼓励多元主体参与生态治理，重点在于不断提高民众的生态环境意识，让民众正确把握生态治理的方向，形成生态文明建设的共识，促进多元化主体积极参与、协同共治。

Chapter 1
Comprehensively Promote the Modernization of State Governance System and Governance Capabilities in the Field of Ecology & Environment

The 18th National Congress of the Communist Party of China (CPC) incorporated ecological civilization into the five-in-one overall agenda for the building of socialism with Chinese characteristics. The Third Plenary Session of the 18th CPC Central Committee set the overriding goal of modernizing state governance system and governance capabilities to deepen the reform comprehensively. It means that promoting modernization of state governance system and governance capabilities in the field of ecology and environment has two levels of strategic significance: Firstly, it contributes to improving ecological environment and promoting the coordinated development of the social and economic system and the ecological system, thus realizing the harmonious co-existence between man and nature. Secondly, it is conducive to modernizing state governance system and governance capabilities in the field of ecological civilization.

"We should persevere in pursuing our lofty ideals so as to achieve long-lasting success." For China, achieving ecological progress is both an opportunity and a challenge. A series of highly effective methods should be adopted to seize this opportunity and overcome the challenges. Promoting the modernization of ecological governance is the inevitable road to ecological civilization construction. At present, it is urgent to clarify the concept of ecological governance modernization and the pathway to achieving it.

Modernizing Ecological and Environmental Governance in the Context of State Governance Transformation

In the context of contemporary China, the state governance system refers to the system the Chinese Communist Party, which is the ruling party, adopts to lead the Chinese people and effectively govern the country. State governance capabilities refer to the abilities of using the state governance system to effectively manage state affairs. The modernization of the state governance system and governance capabilities means that the state institutional system can evolve and transform with the times to support and adapt to China's socialist modernization drive. Since the reform and opening up in late 1970s, China has achieved sustained high-speed economic growth and it has been the second largest economy in the world since 2010. The economic performance of state governance has been recognized by the Chinese people and has also aroused worldwide attention. However, the cost of rapid GDP growth is the deterioration of ecological quality, which has affected people's quality of life and even their health. Although the voices of protecting the ecological environment have been increasing since the 1990s, the speed of the ecological governance has not kept up with the speed of pollution and environmental damages and the ecological crisis has become more and more serious. It has even posed a threat to the health and even survival of the people. Although there have been increasingly loud voices for ecological protection since the 1990s, ecological crisis has been deteriorating, even posing threat to human beings' existence. China has no intention of following the old path of "treatment after pollution", thus protecting environment in the course of development and securing development while protecting the environment has become one of the most resounding themes of the era. Whether environmental pollution can be controlled in order to achieve ecological progress has become an important indicator of the effectiveness of the governance system.

The purpose of the modernization of the state governance system and governance capabilities is to accelerate the pace of institutional construction and improve the ability of state governance in order to ensure the smooth progress of China's modernization, including the ability to effectively control environmental pollution, implement ecological protection and achieve ecological progress. If pollution is out of control and ecological environment is destroyed, this indicates that there is a deficiency in state governance. That is, there is an institutional dilemma in state governance, or there are serious deficiencies in the state governance system. From a macroscopic perspective, China has been faced with challenges caused by historic transformation in the process of China's modernization. Namely, major structural changes in economic and social relations during the process of modernization have produced a large number of economic and social conflicts. However, these conflicts can not be self-correcting in economic and social areas and the state needs to perform intervention through various governing measures. Therefore, the modernization of state governance system and governance capabilities is the institutional adjustment and improvement in order to resolve these conflicts.

Without any doubt, ecological governance modernization is an important part of state governance modernization. Its connotation can be understood from two aspects, namely, pressure in the real world and reflection on the tradition. Regarding pressure, ecological crisis has become a major threat we face presently. The situation has changed from "People regard sufficient food as their heaven" to "people have to depend on heaven for their food". In the last few years, there have been many ecological incidents related to political, economic and social aspects in different places of China. Ecological problems have even become a time bomb. Without a good environment, everything will be in vain. Regarding reflection on the tradition, the emergence of ecological crisis is the inevitable result of traditional approach of development. If we want to make fundamental differences, we need to change our approach of development and reconsider the relationship between ecological protection and economic development.

Even if China had realized the importance of protecting the ecology at the beginning stage of her development, she, as a developing country, lacked the motivation and capability to carry out the protection, thus she was in a paradox in which she could not protect ecology without development and there would be no development if she protected the ecology. However, if China does not protect the ecology now, there will be zero development or negative development. Therefore, the crisis in the real world and the reflection on the traditional approach of development have led to ecological governance modernization. Ecological crisis is an important part of the overall crisis in the transformation. Accordingly, ecological governance modernization is an important part of the modernization of state governance system and governance capabilities. On the one hand, ecological governance modernization needs to ensure the modernization of governance measures, i.e., these measures should be legal, institutionalized, standardized and diversified. What is more important is that ecological governance modernization inevitably mean the modernization of ecological governance capabilities. All in all, ecological governance should produce tangible effect, resolve ecological crisis, achieve the balance between ecological protection and economic development, ensure the sustainable development of the Chinese nation and contribute our unique strength to solving ecological problems in the world. If the above goals are fulfilled, changes must be made regarding development concepts, legal systems, stakeholder diversification and technical support, etc. The process of the changes is the process of promoting the modernization of the ecological governance system and governance capabilities.

The Roadmap to Achieving Modernization of Ecological and Environmental Governance System and Governance Capabilities

We should establish the concept of green development and regard green innovation as a driving force. Clear water and green mountains are invaluable assets. Ecological governance adheres to modern philosophy on ecology and requires the change of development momentum. We should change the traditional exogenous development model of relying on resources, capital and labor expansion for economic development and stimulate innovation as a new engine for economic growth. In particular, we should bring into play the leading role of green innovation for economic growth. Green innovation should be initiated in terms of concept, pathway and many other aspects. Good ecological environment should be provided to the general public as a public product in order to build a modern China with ecological progress. Rigorous innovation can improve the quality and efficiency of economic operation, nurture new economic growth points and boost economic development through developing green products, introducing and creating green technologies, exploring green consumption markets, adopting green resources, upgrading green management systems and building green financing systems, etc.

A law-based ecological and environmental governance system should be constructed. Laws and regulations are vital to the governance of a country and good laws are prerequisite to good governance. A sound legal system can clarify the responsibilities of ecological governance stakeholders, improve recognition of ecological governance and ensure the smooth operation of the ecological governance system. Therefore, it is necessary to build a property right system for ecological environment, enhance the legislation in ecological and environmental protection, natural resource preservation, pollution control, energy security and other important fields and refine the laws on ecological coordination in circular economy, renewable energies and other sectors. Meanwhile, the supervision and management system and accountability system should be perfected to encourage enterprises, people and social organizations to engage in ecological governance so as to establish an all-round and multidimensional ecological governance system.

Constant efforts should be made to enhance ecological governance capabilities. The key to ecological governance modernization lies in the modernization of ecological governance capabilities. The following should be done to improve ecological governance capabilities: Firstly, the Party's leadership should be strengthened. The government boasts the advantage of concentrating resources on major projects. The Party and the government shoulder the same responsibilities and need to work together. Those who neglect their duties shall be held accountable. Supervision and monitoring mechanism should be introduced in the governance process to ensure the governance measures per se are in line with relevant laws and regulations. Secondly, the competition among local governments should be guided in the direction of promoting ecological civilization. Ecological advantages can be turned into true competitiveness in economic development in different regions through improving the ratio of local governments' revenues to expenditure, introducing ecological performance assessment and implementing the uniform environmental standard within the region. In this way, ecological governance becomes the regular and normal work of local governments instead of emergency response to ensure that ecological governance can be carried out on a continuous basis. Thirdly, law enforcement capabilities should be enhanced to promote green development. The series of laws on ecology could only be effective through strict enforcement. This means not only everyone should perform their duties according to law, but also violators shall be punished. Law enforcement capability determines the real effect of governance. Therefore, enhancing law enforcement is of irreplaceable and special significance for ecological governance modernization. Fourthly, a set of comprehensive and sophisticated ecological evaluation indicators should be developed and third-party assessment should be conducted on the results of ecological governance. The enhancement of ecological governance capabilities should lead to the improvement of ecological environment which is the final touchstone of the governance process. Therefore, ecological governance should be result-oriented. The third-party think tanks' role should be brought into play because third-party evaluation can play a unique role in ecological governance. Their engagement can ensure the process of ecological governance is improved and that ecological governance capabilities enhanced.

Thorough integration of ecological governance practices with science and technology should be promoted. As the saying goes, a handy tool makes a handy man. Technology is the powerful tool for transforming nature. Ecological governance modernization cannot be realized without the support of science and technology. We need to promote scientific research in the field of ecological governance, guide governance practice with proven research findings and adhere to the principle that all ecological governance should be based on the practical conditions. In this way, we can prevent haphazard governance and achieve more with less. Therefore, we need to integrate science and technology with ecological governance so as to enhance our ability to transform nature and achieve coordinated development with nature. The innovation in and integration of energy and communication technology will play a decisive role in the process of economic transformation of the human race. At present, we are in a new era in which renewable energy and the information technology are being integrated. Vigorously developing wind power, solar energy, biomass energy and other renewable energy by making use of the big data, cloud computing, "Internet plus" and other smart technologies and improving the efficiency of energy systems in our society are of great significance to ecological governance modernization. They are also critical forces in promoting economic transformation and upgrading in the long run.

Coordinated governance by diversified stakeholders should be promoted. Stakeholders in ecological governance include the state, enterprises, the public and social organizations, etc. The state should pay a leading role by formulating governance policies and ensuring policy implementation. Meanwhile, enterprises, citizens and social organizations should be guided to participate in decision making, supervise and monitor the governance process and make objective evaluations on governance performance. Successful experience and historical lessons of developed countries prove that it is impossible to achieve success if we just rely on government to conduct ecological governance. Participatory governance by diversified stakeholders has become the broad social consensus, and everyone should be encouraged to get involved in ecological governance with the focus on improving the general public's ecological awareness so that the public can understand the direction of ecological governance and reach consensus on ecological civilization. Coordinated governance by all stakeholders should be promoted.

第二章　生态治理现代化的思想基础和理论共识

地理大发现开启了人类全球化的历史，改变了各民族历史相互隔绝的状况，使人类几大文明在发展演进中相互遭遇并走向融合。东西方文明生态观的藩篱古已有之，正是由于地理大发现，才使得东西方生态观得以相互交融。因此，理解生态观的思想基础和理论共识，就需要以地理大发现为节点，认识东西方古代各自的生态观，分析生态观从古至今的变化，并在此基础上形成共识。

从古至今，人类关于自然的观念发生了重大的变迁。在古代，人类将自然看作是养育者，对自然倍加呵护、倍加关爱；近代，随着科学技术的发展和资本主义的崛起，人类开始改变这种有机的自然观，逐渐走向了机械论自然观，开始对自然进行征服，由此导致了今天严重的生态危机；到了近代，随着生态危机的日益严重，人类开始重新认识自然，重新思考人与自然的关系。生态运动的兴起是对人类活动反思的结果，生态治理现代化就是要重新复活活生生的自然，重建人与自然的关系。在一定程度上，这是对古代生态观的复归。这种生态观的变化，反映的不仅是人对自然的重新认识，更是对人类自身命运的思考。因此，重新思考和发掘这些思想的现代价值，是推进生态治理现代化的必然要求。

西方的有机论自然观

无论是古代东方还是西方，在对待自然的观点上毫无疑问是相通的，他们都把自然看作是人类的养育者，以母亲的形象来做隐喻，对自然的利用首先是以对自然的保护为前提的，这实际上是一种有机论的自然观。

从西方来看，不论是古代的哲学、文学、绘画艺术，还是炼金术和采矿业，都将自然尤其是地球与一位养育众生的母亲相等同。她是一位仁慈而善良的女性，在一个有秩序的宇宙中给人类提供所需要的一切。这种有机论的自然观，隐含着一种道德上的规范性约束，当人类将地球看作母亲时，自然会联想到人类自己的母亲，这就从社会道德层面限制了人们对待地球的行动类型，即更多地给予地球以保护，而不是征服。例如，当时的炼金术认为，金属是在地球子宫内成长并转变成黄金的，这就像孩子在女性子宫的温暖窝巢中孕育成长一样，同样，地球的泉水就像是人体的血液系统，地球以人体的样子进行运转，对地球的贪婪开发就是搜寻地球母亲的全身，这在道德上是耻辱的、被禁止的。这种有机论的自然观，在近代以前一直占据着主导地位，主张人与自然之间是和谐相处的关系。自然是人类生存繁衍的基本条件，因此，要求人类珍惜生存条件，努力保护自然生态的平衡，与自然和谐相处，这实际上也是一种整体性的生态观。

中国古代的生态自然观

古代中国作为四大文明古国之一，在与自然和谐相处的过程中，孕育出了灿烂的文化，创造了中国独特的文明奇迹。中国传统文化中蕴含着丰富的生态保护的思想，并且也是以母亲的形象作为隐喻的。例如，道家曾提出自然无为和天地父母的思想，从"人法地，地法天，天法道，道法自然"到"一生天地，然后天下有始，故以为天下母，既得天地为天下母，乃知万物皆为子也，既知其子，而复守其母，则子全矣"，这就是说，天地万物同人的关系，如同父母和子女的关系；再如"穿地皆下得水，水乃地之血脉也，今穿其身，得其血脉，宁疾不邪"，所有这些都反映出有机论的生态自然观，其中所隐含的道德比喻，实际上是对人的行为的一种限制。儒家思想中也包含了生态观念，荀子尽管以"制天命而用之"为人所知，但是他所提出的以时休养生息、保护自然资源的观点，毫无疑问是古代最为杰出的生态思想，如他说"春耕夏耘秋收冬藏，四者不失时，故五谷不绝，而百姓有余食也……斩伐养长不失其时，故山林不童，而百姓有余材也"，这是农业社会进行生产的基本原则，也是与那个时代相适应的生态保护规则。

此外，佛教思想中也蕴含着丰富的生态观念，如禅宗的"郁郁黄花，无非般若，青青翠竹，皆是法身，一花一世界，一叶一菩提"就是无情有性思想的生动体现，这就是说，即使是山川草木等无情物也有佛性，自然万物都是佛性的体现，不仅具有工具价值，更具有自身的内在价值，这与近代的生态观已经相当接近。唐宋以来，儒释道三家逐渐走向融合，至宋代孕育出了理学，而理学本身则继承了三家精华，生态观上也毫不例外，如理学家程颐、程颢对于天地自然之理就曾说道，"天地之间，非独人为至灵，自家心便是草木鸟兽之心也，但人受天地之中以生尔"，主张以己心体草木鸟兽之心，饱含着对动植物的"同情"。至明朝，以王阳明为代表，生态观依然是整体性的和有机的，如王阳明就在《大学问》中提出了"万物一体"的生态观。所有这些，都是古代中国的智慧，对于今天的生态治理具有重要的启示意义。

马克思主义的生态观

近代以来，随着科学技术的发展和资本主义的强势崛起，人类对自然的观念也发生了转变。作为理性的另一面，自然的无序、混乱和灾难逐渐成为人类克服的缺点，以"知识就是力量"为代表的科学革命则使得自然不再神秘，并赋予人类以力量，借以征服自然。资本主义推动生产力巨大发展的同时，使得资本的逻辑得以无限度地扩张，自然资源作为地球孕育的产物，不再具有道德的含义，而只是成为获得利润的原料。在资本的增值面前，人类开始失去理智，疯狂地对自然进行掠夺。伴随着资本主义的全球扩张，生态问题也成为一个全球性问题。马克思指出，"资产阶级在它的不到一百年的阶级统治中所创造的生产力，比过去一切世代创造的全部生产力还要多、还要大"，同时，资本主义在对自然的破坏方面也超过了以往时代的总和，生态危机已经成为我们生存的首要威胁，那个活生生的具有生命力的自然已经被人类杀死了。

马克思主义对资本主义生产方式带来的恶果进行了深刻的批判。马克思主义蕴含着丰富的生态思想，如马克思在《1844年经济学哲学手稿》中就指出,异化的解决要到共产主义社会才能实现,共产主义作为完成了的自然主义,就等于人道主义,而作为完成了的人道主义,就等于自然主义,对自然十分重视。在《德意志意识形态》中,马克思、恩格斯指出:"自然界,就它本身不是人的身体而言,是人的无机的身体。"这是一种辩证地看待人与自然关系的观点。一方面,自然界不是人的身体,人只有通过实践才能认识自然、改造自然;另一方面,自然又是人类的无机身体,人类依赖自然而生存,对自然的破坏相当于毁灭人类自己。因此,马克思主义的生态观可以总结为人在与自然和谐的前提下,以实践为基础认识和改造自然。20世纪以来,生态学马克思主义继承和发展了马克思的观点,认为资本主义制度和生产方式是生态危机的根本原因,有一定的借鉴意义。

构建中国特色的社会主义生态观

20世纪60年代以来,生态运动日渐兴起,生态治理日益成为人类的共识。生态治理是人类为救赎自身而向自然的赎罪,这既是自然的复活之路,更是人类的复活之路。实现中华民族的伟大复兴,生态文明建设是必然之路。社会主义生态文明建设,推进生态环境领域国家治理体系和治理能力现代化,要充分吸收和利用一切人类关于生态保护的优秀思想,做好中国传统文化中生态观的创造性转化和创新性发展,更重要的是要采用马克思主义的基本方法,结合我国实际,构建中国特色的社会主义生态观。

中国特色社会主义生态观要求改变发展理念,不再依靠传统的工业化模式进行发展,树立绿色发展理念;要求全社会树立人与自然休戚与共、和谐共生的意识,凝聚为整个社会的共识,在此基础上促进民众将中国特色社会主义生态观内化于心、外化于行,更好地推进生态文明建设。

Chapter 2
The Ideological Basis and Theoretical Consensus on Ecological Governance Modernization

The great geographical discovery initiated human history of globalization and changed the mutually secluding state of nationalities so that world civilizations could encounter each other and became integrated. The differences between east and west perspectives on ecology have existed since ancient times. It is due to the geographical discoveries that these perspectives could mingle with each other. Therefore, if we want to understand the ideological basis and theoretical consensus on the outlook on ecology, we need to take the great geographical discovery as a starting point so as to understand the east and west perspectives on ecology, analyze the changes since ancient times and reach consensus on this basis.

The concept of human beings on nature has drastically changed since the beginning of human history. In ancient times. People regarded nature as nurturer, so they showed great respect and care for nature. However, with the development of science and technology and the rise of capitalism in modern times, people began to change this kind of organic outlook on nature and gradually adopted Mechanism, so they started to conquer nature, which led to serious ecological crisis. When ecological crisis is increasingly grave, people began to have a new understanding of nature and rethink the relationship between man and nature. The rise of ecological movement is the result of reflection on human activities. Ecological governance modernization is to revitalize the animated nature and reconstruct the relationship between man and nature. To a certain degree, this is the return of ancient outlook on governance. The change of ecological outlook reflects not only people's new understanding of nature, but also the reflection on the destiny of Mankind. Therefore, rethinking and exploring the modern value of these concepts is the inevitable requirement for promoting ecological governance modernization.

The Organicism Outlook on Nature in the West

Whether in the east or in the west, people have similar views on nature without any doubt. They regard nature as the nurturer of the Mankind and called it Mother Nature in metaphor. Protection is a prerequisite for utilizing nature, which is an organicism outlook on nature.

In the west, whether the ancient philosophy, literature and painting or alchemy and mining industry all regard nature and especially earth as a mother, a merciful and kind lady who nurtures all living creatures and provides everything human beings need in an orderly universe. This outlook on nature implies a moral normative constraint. When the Mankind regards the earth as Mother, naturally this will remind them of their own mother, which can restrict people's types of behavior in treating the earth at a moral level. That is, human beings need to protect, rather than conquer, the earth. For example, alchemy in ancient times held that metals grew up in the wombs of the earth and turned into gold in the same way a fetus grows up in the warm nest of a female womb. Likewise, water on the earth is like the blood system of the Mankind. The earth operates like a human body and the greedy exploitation of it is like searching the body of Mother, which is morally shameful and should be forbidden. This kind of outlook on Nature advocating a harmonious relationship between Man and Nature had been occupying a dominant position before modern times. Nature is the basic conditions for human beings to survive and multiply. Therefore, it is actually a holistic outlook on ecology to require human beings to cherish the living conditions, work hard to maintain the balance of ecology and coexist with nature harmoniously.

The Ancient Outlook on Nature and Ecology in China

As one of the four oldest civilizations in the world, China has created a splendid culture with unique miracles in the process of harmonious coexistence with nature. The traditional Chinese culture is abundant in concepts of ecological protection. The metaphor of the earth as Mother is also used. For example, Taoism proposed the idea of actionless activity and treating earth and heaven like parents. It says, "man follows the laws of the Earth and the Earth follows the laws of Heaven, Heaven follows the laws of Dao and Dao follows the laws of Nature" and "Dao is the way of the Earth and Heaven. If we follow Dao, we will never fail". That is to say, the relationship between Nature and human beings is like that between parents and children. Also mentioned by another Taoism book, "Water is the blood of the Earth. If you obtain its blood by digging into its body, then the Earth will become sick". Also these sayings reflect the organic outlook on nature. The metaphor hidden in them is in fact an restraint of human behaviors. Confucianism also contains concepts on ecology. Although Xun Zi, a famous Confucius scholar, is well-known for his idea of Utilizing nature by conquering it, the viewpoint he held that natural rehabilitation in due time and protecting natural resources are the best ideas on ecology in ancient times. He also mentioned "People should plough and sow seeds in spring, kill weeds in summer, harvest in autumn and store in winter. All these things should be done at the right time, so the crops will produce enough yields and the people have surplus food. The logging and planting of trees should also be done at the right season, so the mountains will not become bare and the people will have sufficient timber." This is the fundamental principle of production in agricultural society and also the rule for ecological protection of his era.

In addition, Buddhism is also rich in concepts of ecological protection. In Zen, it says "see the world in a flower and find enlightenment in a leaf". That is to say, even inanimate things like mountains, rivers, grass and trees, contain Buddhism enlightenment. Everything in nature are embodiment of Buddhism. They have their own inherent values in addition to being used as practical tools, which is quite similar to outlook on ecology in modern times. Starting from Tang and Song Dynasties, Buddhism, Confucianism and Taoism were gradually integrated and Neo-Confucianism came into being as a result in Song Dynasty. Neo-Confucianism inherited the essence of the three schools including their ideas on ecology. For example, two New-Confucianists named Cheng Yi and Cheng Hao once commented, "In the planet, Mankind is not the supreme creature and flowers, forests, birds and animals are all equal beings". He maintained that human beings should try to understand other species in the world and be sympathetic to them. In the Mining Dynasty, people held the integrated and organic outlook on ecology with Wang Yangming as a representative. For example, Wang Yangming proposed his outlook on ecology that all things on earth constitute an organic whole. All these reflect the wisdom of ancient Chinese people, which are inspiring to ecological governance today.

Marxism's Outlook on Ecology

The mankind's concept on nature has undergone great changes with the development of science and technology and the rise of capitalism in modern times. As creatures of reason, human beings have tried to overcome the shortcomings such as disorder, chaos and disasters in Nature. Revolution caused by science represented by Knowledge is power makes nature not as mysterious as in the past. Science also empowers human beings to conquer nature. While promoting the rapid development of productivity, capitalism has also expanded in an unlimited manner. As products of Mother Nature, natural resources have lost the moral implication to become only raw materials for making profit. Man began to lose their mind at the temptation of the value created by capital and started to exploit nature in a crazy manner. With the global expansion of capitalism, the issue of ecology has become a global one. As Marx pointed out, the productivity created by the bourgeoisie in less than 100 years is even greater and larger than all the productivity created in all the previous generations. Meanwhile, the damages caused by capitalism are also greater than the sum of all damages in previous eras. Ecological crisis has become the primary threat to our survival and nature with vitality has already been killed by Man.

Marxism has profoundly criticized the evil consequences caused by the capitalist mode of production. Marxism contains a wealth of ecological ideas. For example, Marx pointed out in his *Economic & Philosophical Manuscripts of 1844*, these abnormal issues will only be resolved in communist society. Marxism equals humanitarianism and naturalism and attaches great importance to Nature. Marx and Engels pointed out in *The German Ideology*, nature in itself is not a human body. Rather it is the inorganic body of Man. This is a dialectical view of the relationship between man and nature. On the one hand, Nature is not a human body. Human beings can only understand and transform Nature through practice. On the other hand, Nature is the inorganic body of Man. Human beings' survival depends on Nature, so damaging Nature is equivalent to destroying human beings themselves. Therefore, Outlook on ecology in Marxism can be summarized as "understanding and transform Nature on the basis of practices under the precondition of harmonious co-existence between Man and Nature. Since the 20th century, Ecological Marxism has inherited and developed Marx's points of view and believes that the capitalist system and its mode of production are the root cause of ecological crisis, which is quite enlightening.

Developing Socialist Outlook on Ecology with Chinese Characteristics

Starting from 1960s, ecological movement has been very popular and ecological governance has become consensus of mankind. Ecological governance is the human beings' atonement for their misbehaviors to Nature. This is not only road of resurrection of nature, but also of mankind. Ecological civilization construction is an inevitable road to achieving the great rejuvenation of the Chinese nation. In order to promote modernization of national governance system and governance capabilities in the field of ecology and environment, we need to fully absorb and draw on all the excellent ideas of the world on ecological protection, creatively transform and develop the outlook on ecology in traditional Chinese culture. Furthermore, we should develop socialism with Chinese characteristics by integrating the basic methodology of Marxism with Chinese realities.

The socialist outlook on ecology with Chinese characteristics requires that concept on development be changed. We should establish a green development concept rather than rely on the traditional mode of industrialization for development. The whole society should raise awareness and build consensus on harmonious co-existence between man and nature. On this basis, the public can remember the socialist outlook on ecology and put it into practice so as to promote ecological progress.

第三章　生态治理需要强化法治思维

法治是安邦固本的基石，是治国理政的基本方式，是国家治理体系和治理能力现代化的必要条件和重要特征。建设生态文明、实现绿色发展、推进国家生态环境治理体系和治理能力现代化是一场涉及生产方式、生活方式、思维方式和价值观念的革命性变革，要实现这样的变革，必须善于运用法治思维和法治方式来构建健全的法律制度体系，推进体制机制变革，提高执政能力和水平，建立公众对于法律的信仰。"生态法治化"既是依法治国的必然要求，也是加强生态文明建设的必然选择。

法治的合法与正当价值有助于达成生态治理共识

法治是现代文明国家的标志，法治是规则之治，要求实行良法下的善治，即"已成立的法律获得普遍的服从，而大家所服从的法律又应该是本身制定良好的法律"。从立法来看，法律制定经提出草案、征求意见、审议、表决等程序，是广泛汇集民意、凝聚社会共识的过程，促使社会公众采取自觉行动参与治理。

"徒法不足以自行，良法还需良吏。"然而现实中在一些地方，执法不公、乱作为、不作为等问题仍然存在，懒政、怠政、暴力执法等治理方式往往会引起人们的质疑，甚至反抗，从而良法不行，治理迟滞。2016年1月，最高人民检察院发布了15起检察机关提起公益诉讼典型案例中，有7个地方环保局和3个地方国土局因怠于履职被通报。2016年中央第一批环保督查8省掀起问责风暴，已有近千人被问责，可见恶行并非无存，良吏尚需培养，完善执法人员素质，对实现法治的正当价值至关重要。

公正是法治的生命线。只有依靠法治方式化解矛盾，解决利益冲突，构建理性、和谐的社会环境，才能使社会成员自觉守法，从而形成法律信仰。

民众对政府治理行为的认同、支持、服从，并不是出于功利主义的阳奉阴违，也不是由于"明知不对，少说为佳""明哲保身，但求无过"的消极服从，而是出于对政府生态法治的合法性和公正性的内心确信。

法治的规范和可预期属性有助于明确生态治理旨向

"法律不明确，等于无法律。"规范性是法律语言的基本追求。法律的规范性载负着法的基本的价值，包括自由、秩序、效率等。相较于各种政策文件，法律既能够准确体现国家有关生态治理的旨向和精神，消弭理解上的各种分歧，又能够清楚划定生态治理过程中各个主体的权力（权利）边界，切实把"权力关在笼子里"，从而保障公民的权利。应该按照下位法必须符合上位法的原则，对生态治理的规章和规范性文件进行"立改废"。

规范性文件具有反复适用性和普遍约束力，是各级政府实施行政管理的重要抓手。但是，当前一些政府部门法治观念淡薄，制发规范性文件程序不明、要求不严，导致"红头文件"泛滥，内容不规范、无力监管、政策无法落地。必须以法治方式规范"红头文件"，加强备案审查制度，把规范性文件纳入备案审查范围，依法撤销和纠正违法的规范性文件。严格规范性文件的制定程序，在制定规范性文件时应广泛听取各方面意见，增加透明度。提高"红头文件"的执行效率，做到"无事不发文，发文必落实"。

合理的预期来自于法治。法治使人们在依照法定的"游戏规则"行事的同时，对结果有一定的预期。要不断总结生态治理的经验，强化法治思维，把一些成熟的做法和具有普遍适用性的习惯规则上升为法律规范，形成一体遵行的效力。

法治的整合与协调作用有助于提升政府治理水平

生态治理失灵往往是由于政府部门间职责不清、地区间利益不均等因素造成。当前从中央到地方，具有生态治理职责的部门很多，分散在发改委、环保、国土、工信、住建、农业、林业、水利、交通、海洋等多个部门。在实践中，生态治理涉及多个部门，而各部门间权责不明确、交叉乃至冲突，加上跨区域、跨流域等"条"和"块"不对称因素，经常出现"多龙治水""多龙治海""海洋部门不上岸、环保部门不下海"的现象。生态环境作为公共物品，具有很强的外部性，地方利益分割制约了系统性、整体性的生态治理体系构建和落地实施。

法治的本质是对权力的规范和对权利的保护。有效的生态治理需要用法律的形式明确划定各部门的职责，以此作为治理的依据，避免职责交叉而相互推卸责任；建立体现生态文明要求的绩效考核体系，使得责任意识强化、治理积极性提升；完善部门协作机制，加大沟通协调能力，实现山水林田湖共同体治理。允许政府基于法治理念创新治理方式并给予自由度，有效弥补传统管理模式的不足，实现"带着镣铐跳舞"。

法治的救济和惩罚功能有助于环境修复和生态治理

"有权利必有救济""有损害必有赔偿"是法治社会的基本准则。生态治理法治化的重要功能就在于对环境侵害进行有效的救济和补偿。通过建立合理的环境损害赔偿机制，由造成生态环境损害的责任者承担赔偿责任，确保污染受害者获得合理赔偿，从而责任共担，修复受损生态环境，有助于破除"企业污染、群众受害、政府买单"的困局。需要从立法上明确规定生态环境损害的赔偿范围、责任主体、索赔主体、索赔途径、损害鉴定评估机构和管理规范、损害赔偿资金核定等基本问题。

法律的强制性和权威性，使其具有一般性政策所不具有的威慑力。新修订的《中华人民共和国环境保护法》（以下简称《环境保护法》）赋予环保部门按日计罚、查封扣押、限制生产、停产整治等措施和手段，被称为"史上最严环保法"。建立生态环境法治"高压线"，强化责任追究力度，对破坏生态环境行为的惩罚和制裁会起到震慑效果，从源头上有效减少生态破坏。

法治的民主与开放特性有助于公众参与生态治理

民主参与是法治的生命。法治的权威性是公众参与的保障,而民主与开放的特性又为其注入了活力。权利主张是公众参与政治的基本环节和重要内容,生态环境与每个人息息相关,随着民众权利意识的高涨,参与决策的要求异常强烈。近年来,各地接连发生的反 PX 项目、高铁线路、垃圾填埋场等的环保类群体性事件。"欢迎建设,但请远离我家后院(Not In My Back Yard)","邻避运动"频频发生,根本原因在于缺乏有效的沟通机制,缺乏发育成熟的社会组织运作系统,缺乏合理的利益补偿机制。

化解"邻避"困局,必须运用法治思维和法治方式,提高地方政府生态治理能力,建立多元化的沟通平台,从而形成有效的沟通机制,探索合理的利益补偿方式,使不同利益主体求同存异,依法实现自身利益最大化。创新性提升政务公开化程度,完善治理过程的公示,力求每一个治理阶段都实现有效沟通,从而化解老百姓对政府的"信任危机"。社会组织是"邻避运动"的预警者、公众权益的代言人,具有天然的社会公信力与媒体影响力。社会组织要合理引导公众表达利益诉求,并通过制度化的渠道和政府沟通,充分发挥"政民桥梁"的作用,力求把矛盾化解在萌芽状态,避免"小不满而乱大谋"——因公众不满情绪逐渐累积而诱发群体过激行为。积极探索合理的利益补偿机制,本着"谁受益、谁付费,谁受损、谁受偿"的原则,让受损的少数人获得形式多样、"授之以渔"的合理补偿,努力通过补偿机制平衡各方利益关系,进而助推生态治理的全过程。

Chapter 3
Strengthen the Rule of Law in Ecological Governance

The rule of law is the cornerstone of national peace and security, the basic way of governing the country and the necessary condition and important feature required by a national and modern governance system. Building ecological civilization, realizing green development and promoting the modernization of a state ecological and environmental governance system and governance capability involve revolutionary changes in production method, lifestyle, mindset and values. To achieve such revolutions, we must apply the mindset and practices of rule of law to build a sound legal system, promote institutional reforms, improve the capacity building of the government and establish public belief in the law. "Ecological governance by rule of law" is not only an inevitable part of governing the country according to law, but also a necessary choice to strengthen ecological civilization construction.

Legitimacy and Value of Justice Contribute to the Consensus on Ecological Governance

The rule of law is a symbol for modern civilized countries, which requires fair rules and appropriate implementation of good laws, that is to say, "the existing laws are commonly obeyed and the laws that people abide by are in themselves good laws". From the view of legislation, the making of laws comprises drafting, consultation, deliberation, voting and other procedures, which is a process of collecting public opinion and reaching social consensus, to encourage the public to actively engage in.

"Laws cannot play a part on their own and that's why good laws also need good officials to implement". However, in some places, issues in law enforcement such as injustice, disorder, and inaction still exist. Sluggish, slack and violent law enforcement will entail questions and even resistance from the public. Under such circumstances, good laws are meaningless and governance is stagnant. In January 2016, the Supreme People's Procuratorate reported 15 typical public interest litigations, out of which 7 cases are related to malpractice of local Bureau of Environmental Protection and 3 cases to that of local Land and Resources Bureau. In 2016, the first round of inspection on environmental protection in 8 provinces by central authorities set off a storm, with nearly a thousand people being held accountable. It is thus clear that wrongdoing exists and good officials are still needed. It is essential for the value of justice to improve the quality of law enforcement officers.

Justice is the lifeline of rule of law. Only by relying on the rule of law to resolve contradictions and conflicts of interest and to build a rational and harmonious social environment can we make social members abide by the law and have faith in the law. Public recognition, support, and respect of official governance do not stem from utilitarianism or passive obedience to "keep one's breath to cool one's own porridge" or "keep out of trouble", but from confidence in the legitimacy and impartiality of ecological rule of law.

Standardization and Predictability of the Rule of Law Help Define the Purposes of Ecological Governance

"Unclear laws mean no laws". Standardization is the basic requirement of legal language. The normative nature of the law embraces the basic values of law, including freedom, order, efficiency, and so on. Compared with various policy documents, laws can not only accurately reflect the purpose and spirit of national ecological governance and eliminate various misunderstanding, but also clarify the power (rights) limits of each subject in ecological governance so as to protect the rights of citizens by "confining the power to the cage". Therefore, on the principle that the lower-level law must comply with the upper-level law, rules and regulations of ecological governance should be "newly made, modified, or invalidated".

Regulatory documents are applicable repeatedly and general binding. They are favorable tools for governments at all levels to use in administration. However, some governments think little of the rule of law, with vague procedures of and loose requirements on the making and releasing of regulatory documents. As a result, in these places, "red tapes" are rampage; legal content is not normative, with weak monitoring; and policy are not implemented. Therefore, it is necessary to control the "red tape" reasonably, strengthen the back-up and review system in which regulatory documents are incorporated. Those regulatory documents conflicting with the law should be revoked or corrected. Meanwhile, there should be a strict procedure in making these documents, widely collecting views and increasing transparency. The efficiency of "red tape" implementation should be enhanced to ensure that "there is no red tape without cause so that every red tape will be implemented".

Reasonable expectations come from the rule of law. The rule of law allows people to act in accordance with statutory "rules of the game" and have certain expectations of the outcomes at the same time. We must constantly summarize past experience of ecological governance, strengthen the thinking of rule of law, and upgrade some mature practices and customary rules with general applicability into laws and regulations that are legally binding for the whole country.

Integration and Coordination in the Rule of Law Help Improve Governmental Governance

Failure of ecological governance is often due to unclear responsibilities among governmental departments and unequal interests among different regions. Currently, looking from top-down, there are too many branches that have duties and responsibilities in ecological governance, including different departments affiliated to National Development and Reform Commission, Ministry of Environmental Protection, Ministry of Land and Resources, Ministry of Industry and Information Technology, Ministry of Housing and Urban-Rural Development, Ministry of Agriculture, Ministry of Forestry, Ministry of Water Resources, Ministry of Transport, and State Oceanic Administration. In practice, ecological governance involves a number of departments. As the power and duties among these departments are ambiguous, overlapping, and even conflicting, and due to asymmetry of the governance among cross-regional and inter-basin "vertical" and "horizontal" departments, negative phenomena are exposed, such as "governance of the same matter by difference authorities" and "complete separation from oceanic sector and environmental protection sector". Ecological environment, as public goods, has strong externalities. The fragmentation of local interests constrains the establishment and implementation of a systematic and integrated ecological governance system.

The essence of rule of law is to control and protect power and rights. Effective ecological governance requires a clear definition of the responsibilities of various departments as a basis for governance, to avoid buck-passing caused by overlapping of responsibilities. It requires a performance appraisal system in response to ecological civilization to strengthen the sense of duty and enhance the motivation of governance. It also requires the improvement of the mechanism of inter-departmental collaboration to increase communication and coordination and to achieve co-governance combining authorities that manage mountains, rivers, forests and lakes. The government should be allowed to innovate the way of governance and given certain degree of freedom on the basis of rule of law, to effectively compensate for the weaknesses in traditional management model so that "dancing with shackles" can be achieved.

Legal Relief and Punishment Contribute to Environmental Rehabilitation and Ecological Governance

"Rights go along with relief" and "damages go along with compensations" are basic norms in the rule of law. One important function of ecological governance by rule of law lies in the effective relief and compensation for environmental damage. A reasonable mechanism of compensation for environmental damage should be established to ensure that those who cause ecological damage take the responsibility and compensate victims for pollution. By doing so, responsibilities are shared by those involved to redress ecological damage, which helps get out of the predicament where "corporations pollute, the public suffer, and governments pay". It is necessary to clarify the scope of compensation for ecological damage, the compensator, the claimer, the means of claim, the agency and rules to evaluate the damage, and the verification of the value of compensations.

The coerciveness and authority of laws give them the strongest deterrence not seen in general policies. The newly revised *Environmental Protection Law of the People's Republic of China*, known as the "strictest environmental protection law ever in China", entitles environmental protection departments the rights to punish on a daily basis, to seizure and detain, to restrain production, to terminate production and so on. The "cable with high voltage" in ecological governance by rule of law should be established and those who conduct wrongdoing should be held accountable. Meanwhile, punishment and sanctions should be adopted to effectively reduce ecological damages at the very beginning through a deterrent effect.

Democracy and Openness of the Rule of Law Encourage Public Participation

Democratic participation is the lifeline of rule of law. The authority of rule of law safeguard public participation, and democracy and openness inject vitality into the process. The claim of rights is a fundamental and important part of public participation in politics. Ecological environment is closely related to everyone. With rising awareness of rights, people's demand of participation in decision-making is ever stronger. In recent years, we have seen events initiated by people in groups with same environmental demands, including events against P-Xylene projects, high-speed railway lines, landfills and other projects. "You're welcome to construct but not in my backyard". Activities propelled by such "Not In My Back Yard" (NIMBY) thinking occur frequently. The root cause is the lack of effective communication mechanism, the lack of mature operating system for social organizations, and the lack of reasonable compensation mechanism.

To resolve the dilemma of "NIMBY", we must apply the mindset and means of rule of law, improve ecological governance by local government, and establish a diverse communication platform. By doing so, we can form an effective communication mechanism, explore a reasonable way of compensation, and seek common grounds while reserving differences among different stakeholders to maximize their interests. Government affairs and governance processes should be more transparent and open to strive to achieve effective communication in each stage, so as to resolve the public "crisis of confidence" in the government. Social organizations provide an early warning about "NIMBY" events and speak for public interest, with natural social credibility and influence on media. Social organizations should rationally guide the public to express their interests and demands and act as a bridge between the government and the public through institutional channels to communicate with the government. They should strive to resolve contradictions in the every early stage to avoid deterioration which could lead to radical activities of a group of people propelled by the accumulation of discontent. In the meantime, we should actively explore a reasonable benefits compensation mechanism on the principle that "those who benefit will pay and those who suffer will be compensated". By doing so, the few people who suffer will acquire flexible and reasonable compensations. Through such kind of compensation mechanism, interests of all parties are balanced, which helps enhance the whole process of ecological governance.

第四章 生态治理体系的价值取向和立法路径

"法律即秩序,良好的法律就是良好的秩序。"维持生态环境的良好秩序,必须依靠严格的法律制度。完备的生态环境法律体系是生态治理的前提和保障。改革开放以来,我国在能源、自然资源、环境保护、生态建设等领域出台了系列法律法规,在实施层面也取得了一定的成果;但是,从整体上看,我国生态环境法律体系仍存在着理念滞后、法律缺位、制度缺失、规则冲突、规定模糊等问题,距离生态治理现代化的目标还有一定的距离。以生态文明理念为指导,加快构建科学完善的、以法律为基础的国家生态治理体系,是推进国家生态治理能力现代化的前提保障和必然要求。

健全生态环境法律体系

截至目前,我国出台了多部环境保护相关法律,主要包括《中华人民共和国水污染防治法》(以下简称《水污染防治法》)、《中华人民共和国大气污染防治法》(以下简称《大气污染防治法》)、《中华人民共和国野生动物保护法》(以下简称《野生动物保护法》)、《中华人民共和国水土保持法》(以下简称《水土保持法》)、《中华人民共和国海洋环境保护法》(以下简称《海洋环境保护法》)、《中华人民共和国环境噪声污染防治法》(以下简称《环境噪声污染防治法》)、《中华人民共和国固体废物污染环境防治法》(以下简称《固体废物污染环境防治法》)、《中华人民共和国防沙治沙法》(以下简称《防沙治沙法》)、《中华人民共和国清洁生产促进法》(以下简称《清洁生产促进法》)、《中华人民共和国环境影响评价法》(以下简称《环境影响评价法》)、《中华人民共和国放射性污染防治法》(以下简称《放射性污染防治法》)、《中华人民共和国循环经济促进法》(以下简称《循环经济促进法》)和《环境保护法》等,环境保护法律体系基本形成,各环境要素监管领域已实现基本覆盖。特别是2014年修订的《环境保护法》，进一步明确了政府对环境的监督管理职责，完善了生态保护红线等环境保护基本制度，强化了企业污染防治责任，加大了对环境违法行为的法律制裁，被称为"史上最严"的环境保护法。

然而，当前生态环境法律体系建设依旧存在一些漏洞，例如：部分领域法律缺失，至今仍无法可依。例如土壤污染防治、湿地、核安全、化学品管理、国家公园、生态补偿、环境监测等领域。生态环境保护法律制度缺乏统一协调，之间存在一定程度的交叉、矛盾和冲突，例如区域管理与流域管理之间缺乏协调机制，进而造成管理缺位、效应抵消。许多规定过于笼统、原则模糊，缺少相应配套的法规规章和实施细则，从而难以执行。

提高生态环境立法质量是实现环境法治的前提。应从生态文明建设、绿色发展理念的视角不断健全生态环境法律体系。尽快出台《土壤污染防治法》，确保所有耕地和其他土壤的安全和可持续开发利用。制定《湿地保护条例》，以湿地生态系统或整体流域为主体进行综合管理，明确规定湿地的保护、管理、合理利用以及各方法律责任等。出台《生态补偿法》，按照"保护者得益、受益者补偿、损害者赔偿"的原则来确定生态补偿的方式、种类、范围，完善生态补偿的长效机制。制定《环境监测管理条例》，明确各部门环境监测的地位、性质、职责，以及中央和地方、公益监测和社会监测之间的职责。研究制定《国家公园法》，保障建立国家公园管理体制。制定《动物福利法》，保护动物权益。制定《有毒有害化学物质控制法》。加快修订《循环经济法》《水污染防治法》。以生态文明理念对现行法律法规进行评估，并根据评估结果，对现有生态环境保护法律法规进行清理、修改。通过"亡羊补牢"，周全环境保护法律体系的"竹篱"，与时俱进地"织补"运行系统中的漏洞，努力将现有的生态问题及时控制、潜在的生态危害理性规避、良好的生态环境实现长效合理的保护。

完善自然资源立法

在自然资源领域，我国已构建了以《中华人民共和国宪法》和《中华人民共和国物权法》（以下简称《物权法》）为基础，以《中华人民共和国森林法》（以下简称《森林法》）、《中华人民共和国草原法》（以下简称《草原法》）、《中华人民共和国渔业法》（以下简称《渔业法》）、《中华人民共和国土地管理法》（以下简称《土地管理法》）、《中华人民共和国城镇国有土地使用权出让和转让暂行条例》（以下简称《城镇国有土地使用权出让和转让暂行条例》）、《中华人民共和国城市房地产管理法》（以下简称《城市房地产管理法》）、《中华人民共和国矿产资源法》（以下简称《矿产资源法》）、《中华人民共和国海域使用管理法》（以下简称《海域使用管理法》）、《中华人民共和国水法》（以下简称《水法》）等法律法规为主要架构的法律体系。

当前我国自然资源法律存在的问题主要包括以下几方面。

立法观念陈旧。我国现行自然资源法律大多理念陈旧，甚至带有浓厚的计划经济色彩，保护手段选择上往往采取的是行政强制性义务，缺乏激励性的私法手段，无法调动各方积极性，难以适应社会主义市场经济体制和生态文明建设的新要求。

部门利益保护问题。在自然资源立法方面，我国采取的是针对不同资源，单项立法、部门立法的模式，进而导致了立法部门"多批多获益、少批少获益、不批不获益"的恶性后果，法律在一定程度上成了维护本部门利益的屏障，互无制衡，更无协同。

部分法律制度尚有缺失。如海洋领域长期缺失一部基本性法律。在自然资源产权制度、流转制度、有偿使用制度等存在空白或缺陷的背景下，制度缺失必然导致利益争端、权责不明、资源浪费。根据我国现行规定，所有自然资源归国家和集体所有，但由于缺乏具体的资源产权主体代表，形成众多资源利用、利益分配上的矛盾，导致自然资源利用目光短浅、乱用滥用，加剧了资源配置效率低下、破坏浪费严重。

法律条文欠缺可操作性。自然资源法律的出台由于受立法"宜粗不宜细"的影响，很多条文多为原则性、模糊性规定，欠缺可操作性，容易被"钻空子""捡漏子"，使有关自然资源法律主体的权利、义务等内容不具有确定性和针对性。

应本着因时而动的生态文明理念建立健全自然资源法律体系。加强民法物权理论与自然资源立法研究。重视市场在自然资源配置中的作用，实现自然资源立法的体系化、科学化、生态化。制订自然生态空间统一确权登记办法，健全自然资源资产负债表编制制度，构建归属清晰、权责明确、监管有效的自然资源资产产权制度体系。加快建立健全国土空间开发保护制度和用途管制制度。制定《海洋基本法》，维护我国主权和海洋权益；研究修订《土地管理法》，完善土地权利制度、土地市场制度、征地制度等；修改《矿产资源法》，加强矿产资源集中统一和分级分类管理，建立和完善探矿权、采矿权的权利制度；修改《森林法》，适度放宽林业管制，突出体现分类经营、生态优先等原则。

加快能源法治建设

能源法律体系需要调整的范围包括煤炭、石油、天然气、电力、核能、新能源和可再生能源、能源节约、能源监管等方面。改革开放以来，我国已制定《中华人民共和国电力法》(以下简称《电力法》)、《中华人民共和国煤炭法》(以下简称《煤炭法》)、《中华人民共和国可再生能源法》(以下简称《可再生能源法》)、《中华人民共和国节约能源法》(以下简称《节约能源法》)、《中华人民共和国电力监管条例》(以下简称《电力监管条例》)等，形成了以单行能源法为主，配套法规为补充的立法现状。

从总体上看，我国能源法律体系建设比较滞后，甚至在许多领域长期处于无法可依的状态。当前，我国能源法律体系缺少基础性、综合性的能源法律——《能源法》。同时，在石油、天然气、核电等领域，单行法律缺位。受能源行业特点的影响，《电力法》《煤炭法》等能源法更多体现出部门法、行政法的特征。

能源法治的关键是加快能源法律体系建设，应明确完善我国能源法律体系的路线图和时间表，有计划、按步骤、成体系地加快能源法律领域的"立改废"。

出台"能源领域的基本法"——《能源法》，填补能源领域基础性、综合性法律的空白，使各项能源法律法规以能源基本法为依据、有因循、相协调。研究制定《石油天然气法》等重要领域的能源单行法律。修改《电力法》《电力监管条例》，推进电力市场化改革。加快《原子能法》《核安全法》等法律的制定。完善可再生能源法律体系，细化可再生能源发展目标，丰富监督管理体制和公众参与制度。制定《海洋石油天然气管道保护条例》《国家石油储备条例》《能源监管条例》等行政法规。形成"基本法—单行法"完整统一、有表有里、有粗有细、有筋有骨的法律体系。

Chapter 4
Value Orientation and Legislative Path of Ecological Governance System

"Law is order and good laws mean good order." To maintain a good order in ecological environment, there must be a strict legal system. A full-fledged legal system for ecological environment is the prerequisite and guarantee for ecological governance. Since the reform and opening up, China has introduced a series of laws and regulations in the fields of energy, natural resources, environmental protection and ecological construction and has achieved some results in the implementation level. However, generally speaking, there are still some problems lying behind our country's legal system on ecological environment such as outdated concepts, lack of laws and systems, conflicts of rules and vagueness of provisions. There is still a distance to realizing the modernization of ecological governance. It is the prerequisite, guarantee and requirement to promote the modernization of the national ecological governance under the guidance of ecological civilization concept and to speed up the construction of a scientific, sound and law-based national ecological governance system.

Improving the Legal System in Ecological Protection

Up to now, China has promulgated a number of environmental protection laws including *Water Pollution Prevention and Control Law of the People's Republic of China, Atmospheric Pollution Prevention and Control Law of the People's Republic of China, Wild Animal Conservation Law of the People's Republic of China, Water and Soil Conservation Law of the People's Republic of China, Marine Environment Protection Law of the People's Republic of China, Law on Prevention and Control of Pollution From Environmental Noise of the People's Republic of China, Law on the Prevention and Control of Environmental Pollution Caused by Solid Waste of the People's Republic of China, Law on Desert Prevention and Transformation of the People's Republic of China, Cleaner Production Promotion Law of the People's Republic of China, Law on Environmental Impact Assessment of the People's Republic of China, Law on Prevention and Control of Radioactive Pollution and Circular Economy Promotion Law of the People's Republic of China, Environmental Protection Law of the People's Republic of China*. The basic legal system for environmental protection has formed, and the supervision of environmental elements has achieved basic coverage. In particular, the *Environmental Protection Law*, which was revised in 2014, further clarified the government's responsibility for environmental supervision and management, improved the basic system of environmental protection such as red line of ecological protection, strengthened the responsibility of corporations to prevent and control pollution, and increased the sanctions on breach of environmental law. It is known as the "most stringent" environmental protection law ever.

However, there still exist some loopholes in the current legal systems on ecological environment, for example, lack of laws in some areas, including fields like soil pollution control, wetlands, nuclear safety, chemical management, national parks, ecological compensation, environmental monitoring and so on; lack of coordination in ecological protection laws and overlapping, contradictory or conflicting content to a certain degree. For example, there is lack of coordination mechanism between regional management and river basin management, which leads to the lack of management and offset of effect. Many of the provisions are too general or vague, and are lack of corresponding supporting laws, regulations and detailed implementation, which makes it difficult to implement.

Improving the quality of ecological environment legislation is the prerequisite for realizing law-based environmental regulation. It is necessary to improve the legal system of ecological environment from the perspective of ecological civilization construction and green development concept. An early introduction of *Soil Pollution Prevention and Control Law* can ensure the safety and sustainable development and utilization of all the arable land and other soil. The *Regulations on Wetland Protection* should be developed so that comprehensive management of wetland ecosystems or the overall watershed can be realized. The regulations should clearly stipulate the protection, management, and rational use of wetlands and legal responsibilities of all parties. *Ecological Compensation Law* should also be introduced to determine the way, type and scope of ecological compensation and improve the long-term mechanism according to the principle of "benefit the protectors, compensate the beneficiaries and reimburse the victims". *Regulations on Environmental Monitoring and Management* should be formulated to clarify the status, nature and responsibilities of environmental monitoring in various departments, and the responsibilities between central and local monitoring, as well as public welfare monitoring and social monitoring. It is necessary to research and formulate the *National Park Law* to protect the establishment of the national park management system and enact the *Animal Welfare Law* to protect animal rights, as well as develop *Law on Control of Toxic and Hazardous Substances*. Accelerating the revision of *Circular Economy Law* and *Water Pollution Prevention and Control Law* is of great importance. The government can assess the existing laws and regulations with the concept of ecological civilization, and to clean up and modify the existing ecological and environmental protection laws and regulations in accordance with the assessment results. A "timely remedy" can help reinforce the "fence" of environmental protection legal system. And it is pivotal to "patch" the loopholes in the operating system and keep pace with the times, to control the existing ecological problems in a timely manner and to rationally avoid the potential ecological hazard, and to achieve long-term and reasonable protection of a good ecological Environment.

Strengthening Legislation in Natural Resources

In the field of natural resources, China has built a legal system on the basis of *Constitution* and *Property Law of the People's Republic of China* with the following laws and regulations as the main structure: *Forest Law of the People's Republic of China, Grassland law of the People's Republic of China, Fisheries Law of the People's Republic of China, Land Administration Law of the People's Republic of China, Provisional regulations on the Leasing and Subleasing of Urban State-owned Land of the People's Republic of China, Urban Real Estate Administration Law of the People's Republic of China, Mineral Resources Law of the People's Republic of China, Law on the Administration of Sea Areas of the People's Republic of China, Water Law of the People's Republic of China* and others.

At present, the existing problems of laws on natural resources in our country mainly include the following aspects:

The legislative concept is outdated. Most of the existing laws on natural resources in China are obsolete and some are even full of features of planned economy. The protection means are often administrative compulsory obligation, lacking the incentive private means and can't mobilize the enthusiasm of all parties, thus it is difficult to meet the new requirements of socialist market economy and ecological civilization construction.

There exists protection of departmental interests. Regarding legislation on natural resources, China has adopted a model of department-based legislation for different resources, which has led to the vicious consequences of departments "gaining more benefits for more approvals, fewer benefits for fewer approvals, and no benefits for no approvals". To some extent, the law has become an umbrella to safeguard departmental interests without any check and balance or coordination.

The legal system is incomplete. In the marine field, there has been no basic law for a long time. There are no clear stipulations on natural resources property rights, circulation system and paid-use system. The incomplete legal system inevitably leads to interest dispute, unclear division of power and responsibility and wasted resources. According to the existing provisions of our country, all the natural resources are owned by the state or collectives. However, because of the lack of specific main representatives for property rights of resources, there have been conflicts over resource utilization and benefit distribution, resulting in the short-sighted and indiscriminate use of natural resources, which exacerbates the inefficient allocation of resources, destruction and waste of resources.

Legal provisions are far from operational. The law of natural resources is affected by the legislation principle of "better be general than precise", so many provisions are just about principles and are ambiguous, which can't be operated. They are easy to be picked out with "loopholes" or "missing parts", so contents about legal subject's rights and obligations concerning natural resources are not certain or targeted.

It is necessary to establish a sound legal system on natural resources based on the concept of developing ecological civilization and strengthen the research on the theories on property rights in *Civil Law* and the Legislation on natural resources. The role of market in the allocation of natural resources should be emphasized to make the legislation of natural resources systematical, scientific and ecological. A unified registration system for natural ecological space must be developed to improve the balance sheet formation system of natural resources and build an assets property system for natural resources that has clear ownership, defined rights and responsibilities and effective supervision. It is important to establish and improve the development and protection system of land and space and the usage control systems. The *Basic Law of the Sea* should be formulated to safeguard China's sovereignty and maritime rights and interests. Study and amendment to the *Land Administration Law* must be conducted to improve the land rights system, land market system, and land acquisition system; modify the *Mineral Resources Law* to strengthen the centralized and classified management of mineral resources, and to establish and improve its right of exploration and mining; modify the *Forestry Law*, appropriately relax the forestry control, and highlight the principles of classified operation and ecological priority.

Stepping Up Rule of Law in Energy Field

In the energy law system, the following scope needs to be adjusted: coal, oil, natural gas, electricity, nuclear energy, new energy and renewable energy, energy conservation, energy regulation and so on. Since the reform and opening up, China has developed the *Electric Power Law of the People's Republic of China, Coal Industry Law of the People's Republic of China, Renewable Energy Law of the People's Republic of China, Energy Conservation Law of the People's Republic of China, *Regulation on Electric Power Supervision of the People's Republic of China*, and has formed a legislative status quo focused on special energy law, and supported by supplementary laws.

Generally, the development of China's energy law system is lagging behind, even there are no laws in many areas for a long time. At present, China's energy law system lacks basic and comprehensive energy law – the *Energy Law*. At the same time, in the oil, natural gas, nuclear power and other fields, single line of legislation is in absence. Affected by the characteristics of the energy industry, *Electric Power Law*, *Coal Industry Law* and other energy laws more reflect the characteristics of the department-based law and administrative law.

The key to a law-based energy management is to speed up the construction of energy legal system. There should be a clear road map and timetable of energy law system to make sure the "legislation-amendment-abolition" of energy law can take place in a planned, progressive and systematic way.

As the fundamental law in energy field, the *Energy Law* should be introduced to fill in the blank of fundamental or comprehensive laws, so that all the energy laws and regulations can be based on the Energy Basic Law and be in coordination with it. It is necessary to research and develop specific energy laws such as *Oil and Gas Law*, modify the *Electric Power Law* and *Regulation on Electric Power Supervision* to promote the power market reform. Introduction of *Atomic Energy Law*, *Nuclear Safety Law* and other laws needs to be accelerated. It is important to improve the legal system of renewable energy, refine its development goals, and enrich supervision and management system and public participation system. Administrative regulations such as *Regulations on the Protection of Maritime Oil and Natural Gas Pipelines*, *Regulations on State Petroleum Reserves* and *Regulations on Energy Supervision* need to be developed and enacted. The legal system should be a complete and unified "basic law – special laws" form, which has contents that are both external and internal, general and detailed, with guidelines and implementation.

第五章　创新执法机制　护航绿色发展

实现绿色发展、建设生态文明意味着要在经济建设中彻底摒弃"以资源换发展"的观念，转而在发展中秉持生态、绿色、低碳、循环的理念。绿色发展不仅是破解能源资源和生态环境约束，实现人民群众对绿水青山期盼的必然要求，还是培育新的经济增长点的最佳选择。要将"绿色发展"从口号变成现实，必须在法律法规、制度政策、体制机制上下功夫，方能使绿色发展有法可依，有规可循。

党的十八大以来，我国加快推进资源环境领域法律法规规章的"立改废"，以生态环境保护、资源节约集约利用为主的绿色发展法律体系不断得以完善，加上各类纲要、规划、行动计划等，为实现"绿色法治"奠定了坚实的基础和依据。

"法令行则国治，法令弛则国乱。"法律的生命力在于实施，法律的权威也在于实施。虽然当前绿色发展在很大程度上实现了"有法可依"，但实践中仍存在权责不清、配合缺位、有所掣肘、执法不严、过度执法等诸多问题。因此，必须不断创新执法机制、提升执法能力以护航绿色发展。

构建完善的绿色发展法律体系

有法可依是执法的基础和前提。当前，我国在绿色发展领域已形成一套相对完整的法律体系，基本解决了无法可依的状态。然而，绿色发展法律体系在实践中仍存在"立法有数量无质量，既无大错也无大用"的情况；在污染防治、海洋、湿地、环境监测等领域依然存在法律不完备，甚至立法空白的问题；大量指导性、建议性、鼓励性的法律政策缺乏可操作性、无法落地；部门立法色彩浓重，导致法律受部门利益制约而效力有限。

必须加快完善绿色发展的执法依据。一要健全绿色发展法律体系，在能源、海洋、土壤、湿地等领域尽快填补立法空白，为绿色发展提供有力的依据和保障。二要逐步增强绿色发展法律的体系性和可操作性，清理现行法律中与绿色发展不相适应的内容，修正其中过于笼统的条文，最大可能地将其具体化、明确化，使之从空洞的法条变为实务中便于适用的依据。三要积极推动由部门主导立法向立法机关主导立法的转变，健全由全国人大及其常委会和国务院法制办草拟条文的机制。深入推进科学立法、民主立法、开门立法，拓展社会各界有序参与立法的途径和方式，采取"广纳谏言"式的立法模式。

探索综合执法机制

科学合理的体制机制是有效执法的关键。目前，我国实行部门统管与多部门分管相结合的执法体制，各部门之间的职能分配、责任归属、管辖范围不够明晰，导致部门之间各自为政、互相推诿的现象时有发生，难以形成高效的执法合力。同时，在涉及跨地域、跨流域的问题时，同级同类部门之间常常缺乏协调配合，使得执法效率大大降低。因此，继续修改现行的条块分割体制、建立综合执法机制显得十分必要。整合多部门的职能和资源能为执法提供更坚实的人力、物力、财力保障，使执法效果显著增强。

近年来，从中央到各地都在积极探索综合执法机制。2013年，环保部与公安部联合发文，建立环保和公安联合执法的衔接机制，包括联席会议制度、联络员制度、案件移送机制、重大案件会商和督办制度、紧急案件联合调查机制、案件信息共享机制以及奖惩机制。"环保+公安"双剑出鞘，严惩环境犯罪。浙江舟山亦探索"大海洋"综合行政执法模式，对海洋渔业、港航、国土资源、水利等多个部门的海洋执法职能进行整合，"一个口子对外"综合执法，有效破解了"多龙治海"导致的执法难题。2014年，党的十八届四中全会要求在资源环境、农林水利、海洋渔业等领域内全面推行综合执法。

当下，应当进一步深化绿色发展执法体制改革，健全跨部门综合执法机制。一要理顺部门职责，明晰各部门的职责和权限，避免出现职能冲突或管理空白。二要建立部门间的长效联动机制，探索推进跨部门联合执法、交叉执法，并加强部门间信息交流。三要加强执法刚性，联合打击违法行为，做到违法必究，有案必查，形成高效的执法合力。

健全责权清晰的垂直管理制度

"谁控制了其生存，谁就控制了其意志。"由于现行体制下生态、环保等部门人事和财政受当地政府管理，因此，其在执法过程中往往掣肘于当地政府。要避免地方保护主义，就要割裂地方政府对当地生态环保执法部门"人财物"的控制，确保公正执法、严格执法。

实行省以下环保机构监测监察执法垂直管理制度有助于执法机关摆脱地方保护主义的限制，确保环境监测监察执法的独立性、权威性、有效性。要理顺市县环保部门和环境监察部门之间的关系，避免环境执法缺位、重复执法、过度执法等问题。坚持环境质量属地责任，强化地方党委和政府及其相关部门的环境保护责任。中央环保督察组掀起的治污问责风暴，既是一种治理的示范，也是为"最严环保法"的落地除障清路。应健全长效机制，将整改落实长效化、严格执法常态化。

执法司法联动打出生态保护齐抓共管的"组合拳"

执法与司法之间的断层往往导致执法部门面对环境污染犯罪时单打独斗、势单力薄，使得某些本应受到刑法制裁的犯罪嫌疑人逃脱法网。低犯罪成本与高收益之间的不对等会助长环境污染违法犯罪行为的发生。加强执法与司法的联动能够使环保部门与法院、检察院和公安机关等司法部门形成合力，彻底解决有案不移、有案难移、无案可审和无案可查的局面，有效遏制、打击环境污染违法犯罪行为，从而实现"最严执法"，保障绿色发展。

强化执法与司法联动，打出生态保护齐抓共管的"组合拳"。一要明确行政执法与司法联动的法律依据。二要严格落实违法犯罪线索移送制度，杜绝"以罚代刑"。三要健全案件移送监督制度。明确检察机关监督移送涉嫌环境污染犯罪案件的方式，对于联动过程中可能出现的腐败问题进行监督和严厉打击。四要进一步完善环保公益诉讼，依法支持社会公益组织监督和惩罚企业非法排污。探索设立环保公益基金，将公益诉讼中民事赔偿和刑事案件中的罚金纳入公共性环保专用基金。

加强执法软实力、硬条件综合能力建设

"法之善者仍在有用法之人，苟非其人，徒法而已。"执法者的素质直接决定执法水平的高低。绿色发展领域执法具有较强的专业性，执法人员既熟悉法律法规，又了解资源环境科学知识，才能做到准确应用，精准打击。因此，须加快培养执法人员的工作能力，优化执法团队组成结构。积极开展有关生态学、环境保护、证据采集、科技监管等知识技术培训。组建环境监测、应急、评价等各类专家队伍，设立专职化生态警察队伍，最终建设一支真正懂法律、素质高、业务强、管理精的执法队伍。增加执法人员编制，落实执法经费，提高执法装备质量，避免"案多人少，有心无力"的情况发生。

"工欲善其事，必先利其器。"应当充分利用"互联网＋移动执法"手段，建立国家层面互通共享的移动执法数据库，统一技术标准和数据接口，强化对执法的技术支持。移动执法与大数据、互联网的高度融合，有助于解决跨地区环境污染问题，便于各地执法机关信息交流，避免漏网之鱼；还可借助大数据技术进行量化统计分析，实现全天候、跨地区、多时段的执法模式。通过互联互通和数据共享，实现执法智能高效化、监管精准化、综合决策科学化。努力提高生态环境监测设备的利用效率。在执法人员素质提升的基础上，提高执法设备的科技效用，精确评估生态环境变化情况，是配合"软实力"提升背景下的"硬要求"。

立法评估、监督问责规范立法执法

开展立法评估试点，能够立起一面镜子，为今后法律法规的"立改废"

提供了科学依据，有助于加强执法保障。立法评估有助于立法机关自我反省，及时发现现行法律在执法实践中的可行性问题，提高立法水平，亦有助于社会群体提高法治意识，参与立法过程。建立规范化、程序化、制度化的评估体系。积极推动第三方智库、专家学者与社会公众参与评估工作。

"没有约束的权力必然滋生腐败。"必须加强对执法不严、违法不究的监督和追责，防止执法机关因不正当利益而不作为甚至乱作为。一方面要强化监督，建立行政执法责任制，严格确定并具体落实不同岗位执法人员的执法责任，坚决排除地方和部门保护主义对执法活动的干预。另一方面，要落实问责，建立领导干部任期绿色发展、生态文明建设责任制，实现生态保护"党政同责、一岗双责"，责任具体到人到岗。对环境持续恶化的地方，坚决实行通报、约谈、预警、限批、追责等措施。

Chapter 5
Innovate the Law Enforcement Mechanisms to Safeguard Green Development

To achieve a green development and ecological civilization means the concept of "resource for development" should be completely abandoned in economic construction, but to uphold an ecological, green, low carbon and recycling concept. Green development can not only break the constraints to energy resources and ecological environment, meet people's expectations of green mountains and clear water, but also a best choice to cultivate new economic growth point. To translate "green development" from a slogan into a reality, the government must have achievements in laws and regulations, institutional policies, mechanisms and systems, so that green development will have laws and rules to follow.

Since the 18th National Congress of the Communist Party of China, China has accelerated the "legislation-amendment-abolition" procedure of the laws and regulations in the field of resources and environment. The legal system for green development, which features ecological protection, resource conservation and intensive use, has been continuously improved. Together with other guidelines, plans and action plans, "green rule of law" has been laid a solid foundation and basis.

"A strict law can make sure of a good state governance, while a weak law will lead to state chaos", so the vitality and authority of law lies in its implementation. Currently, green development has realized "There is law to abide by" to a large extent, but in practice, there are still many problems, like unclear definition of rights and responsibilities, absence of cooperation, existence of constraints, weak law enforcement, excessive law enforcement and so on. Therefore, the government must continue to innovate the law enforcement mechanism and enhance its ability to safeguard green development.

Constructing a Full-fledged Legal System on Green Development

Law is the basis and prerequisite for enforcement. At present, China has formed a relatively complete legal system in the field of green development and basically solved the problem of having no laws to resort to in law enforcement. However, in practice, problem still exists in legal system of green development, e.g., "The legislation only has quantity but no quality, and it has no big mistakes or big use"; in fields like pollution prevention, marine, wetland, and environmental monitoring, laws are still incomplete, or even blank; a large number of instructive, suggestive and encouraging legal policies lack operability, so they can't be implemented; some laws are made based on the proposal of government departments, thus only having limited effectiveness because of conflicts of interests.

The government must speed up improving law enforcement basis of green development. First, the government must improve the legal system of green development, fill in the blank of legal system in fields like energy, marine, soil, wetlands and so on, so as to provide strong basis and protection for green development. Second, the government need to make green development more systematic and operational, clean up the contents that are incompatible with green development, amend provisions that are too general, make them more specific and clear, and change them from hollow laws into practical basis. Third, the government should actively promote the transition from the department-leading legislation to the legislature-leading legislation, improve the mechanism where the provisions should be drafted by the NPC and its Standing Committee and Legislative Affairs Office of the State Council, deepen the scientific legislation, democratic legislation, open-door legislation, expand ways and methods with which all walks of life in the society can participate in legislation in an orderly manner, and take legislative model that features "extensively take good advice".

Exploring a Comprehensive Law Enforcement Mechanism

A scientific and reasonable institutional mechanism is the key to effective law enforcement. At present, China implements law enforcement system featuring one department in charge of general management and other departments are assigned with respective responsibilities. Function assignment, responsibility attribution and management scope in different departments are not clear, resulting in the phenomenon that departments administer in their own ways and escape from responsibilities, which makes it difficult to form an efficient law enforcement. At the same time, in the cross-regional and cross-basin issues, similar departments at the same level often lack coordination, making the efficiency of law enforcement greatly reduced. Therefore, it is necessary to revise the existing fragmenting system and establish a comprehensive law enforcement mechanism. Integrating multi-sectoral functions and resources can provide a more solid human, material and financial guarantee for law enforcement and significantly enhance the effectiveness of law enforcement.

In recent years, departments are actively exploring the comprehensive law enforcement mechanism from central to local levels. In 2013, the Ministry of Environmental Protection and the Ministry of Public Security jointly issued an announcement, advocating to establish a joint mechanism which combines the law enforcement of the two Ministries, including the joint meeting system, liaison system, case transfer mechanism, major cases of consultation and supervision system, joint investigation mechanism for emergency cases, sharing mechanism for case information and reward and punishment mechanism. With "Environmental protection + public security" swords combined, environmental crimes will be severely punished. Zhejiang Zhoushan also explores the "big ocean" comprehensive administrative law enforcement model and integrates the marine law enforcement functions in many departments, including marine fisheries, port and shipping, land resources, water conservancy and so on. The comprehensive law enforcement featuring "One hole to outside" effectively solves the problem of "Many dragons govern the ocean". In 2014, the Fourth Plenary Session of the Eighteenth Central Committee of the Communist Party of China called for comprehensive law enforcement in the fields of resource environment, agriculture, forestry, water and marine fisheries.

At present, the government should further deepen reform in law enforcement system in green development, improve the cross-sectoral comprehensive law enforcement mechanism. First, the government need to clarify the responsibilities and authority limits of departments, avoid the emergence of functional conflicts or management gaps. Second, the government must establish a long-term linkage mechanism between departments to explore cross-sectoral joint law enforcement, cross-law enforcement and strengthen inter-sectoral information exchange. Third, the government must strengthen the rigidity of law enforcement, jointly fight against illegal acts, make sure the law-breakers are prosecuted and all cases are checked, and form an efficient law enforcement.

Establishing a Vertical Management System with a Clear Definition of Powers and Responsibilities

"He who controls their survival can control their will." Under the current system, personnel and finance of ecological and environmental departments are governed by local government, therefore, the process of law enforcement is often hindered by local government. To avoid local protectionism, the local government's control of "human and property" on local eco-environmental law enforcement departments should be removed to ensure fair and strict law enforcement.

The implementation of the vertical management of law enforcement in the monitoring of environmental protection agencies under the provincial level will help law enforcement agencies to get rid of restrictions from local protectionism, ensure the independence, authority and effectiveness of the law enforcement in the environmental monitoring. The relationship between the city and county environmental protection departments and environmental monitoring departments should be rationalized. The following issues should be avoided: the lack of environmental law enforcement, repeated law enforcement, excessive law enforcement and other issues. The government should adhere to the territorial responsibility for environmental quality, strengthen the environmental protection responsibility of local party committee, government and relevant departments.

Central Environmental Inspectorate sets off a pollution control accountability campaign, which is a model of governance, and eradicates the barriers for implementing "the most stringent environmental protection law", The government needs to improve a long-term mechanism, make the implementation have long-term effectiveness and strict law enforcement normalized.

Cooperation Between Departments of Law Enforcement and Justice to Protect Ecological Environment

The disconnect between law enforcement and justice often makes law enforcement departments alone and weak when facing environmental pollution crimes, making some criminal suspects who should get criminal sanctions escape from the punishment. Inequality between low crime costs and high returns will contribute to the occurrence of environmental pollution crimes. Strengthening the linkage between law enforcement and justice can make the environmental protection departments and the courts, procuratorates and public security organs form a concerted effort to completely solve the following problems: no pushing forward of the cases, difficulty in pushing forward of the cases, no cases to be judged, and no case to be checked, so as to effectively control and combat environmental pollution crimes, and achieve "the most strict law enforcement" and protect the green development.

Law enforcement departments and judicial departments should work together to protect ecology. First, the legal basis of administrative law enforcement departments working with judicial organs must be clarified. Second, the transfer system of criminal clues must be strictly implemented and administrative punishment should not replace criminal penalty. Third, the supervision system of the case transfer must be improved. The case transfer methods concerning environmental pollution in the procuratorial organs should be specified. The government should monitor and severely crack down on the corruption that may occur in the linkage process. Fourth, the government should further improve litigation of public interest in environmental protection, support social welfare organizations to monitor and punish enterprises' illegal sewage in accordance with the law. The government should explore and establish public fund in environmental protection, and put civil compensation in public interest litigation and fine in criminal cases into the public environmental protection fund.

Strengthening Comprehensive Law Enforcement Capabilities

"Good laws are dependent on the people who enforce laws. If there are no capable people, then the law is useless". The quality of law enforcement officers directly determines the level of law enforcement. Law enforcement in green development is very professional. Only when law enforcement officers are familiar with laws and regulations and the scientific knowledge about resources and environment, can they apply them accurately and crack down on the crimes precisely. Therefore, it is necessary to speed up the training of law enforcement officers and to optimize the composition of law enforcement team. The government need to actively carry out the knowledge and technology training, including ecology, environmental protection, evidence collection, science and technology supervision and so on. A group of experts in environmental monitoring, emergency coping, and evaluation needs to be formed, a full-time ecological police team needs to be established, and ultimately a law enforcement team that really understands the law, with high quality, good business capability, and good management should be built. The related departments need to increase the establishment of law enforcement staff, implement law enforcement funds, improve the quality of law enforcement equipment, and avoid the situation that "cases are more than staff and the staff is powerless".

"A craftsman, if he means to do good work, must first sharpen his tools". The government should make full use of the means "Internet + mobile law enforcement", establish a national database for sharing mobile law enforcement, unify technical standards and data interface, strengthen the technical support for law enforcement. The integration of mobile law enforcement, big data, and the Internet can help solve the problem of cross-regional environmental pollution, facilitate the exchange of information on law enforcement agencies to avoid escape of crimes. Quantitative statistical analysis with large data technology can help achieve all-weather, cross-regional and multi-period law enforcement. Through interconnection and data sharing, we can realize intelligent and efficient law enforcement, precise regulation, and scientific and comprehensive decision-making. Efforts should be made to improve the efficiency of the monitoring equipment in ecological environment. On the basis of improving the quality of law enforcement personnel, the law enforcement equipment should be more scientific and technological so as to accurately assess the ecological environment changes, which is the "rigid requirements" with the background of "soft power".

Standardizing Legislation and Law Enforcement Through Assessment of Legislation and Accountability

To carry out the pilot to assess legislation means setting up a mirror and provide scientific basis of "legislation-amendment-abolishment" for the future laws and regulations. It helps to strengthen law enforcement. The legislative assessment will help the legislature to have self-reflection, timely discover the feasibility of law enforcement in the existing law, improve the level of legislation, and help the community to raise awareness of the rule of law, and participate in the legislative process. A standardized, procedural and institutionalized evaluation system needs to be established. The third party think tank, experts and scholars, and the public should be promoted to participate in the assessment work.

"Absolute power is bound to breed corruption". The government must strengthen supervision and accountability on the lax law enforcement and impunity, prevent law enforcement agencies from inaction or misuse of power because of improper interests. On one hand, the government needs to strengthen supervision, establish a responsibility system on administrative law enforcement, strictly determine and specifically implement law enforcement responsibilities for law enforcement officers in different positions, and resolutely remove intervention in law enforcement activities from local and department protectionism. On the other hand, the government needs to implement the accountability, establish a responsibility system where leaders should promote green development and ecological civilization construction, achieve ecological protection where "party and government share the same responsibility, a post has two responsibilities", responsibilities should be assigned to specific personnel and post. In fields where the environment continues to deteriorate, the following measures should be implemented resolutely: notification, interviews, early warning, limited approval and accountability.

第六章　第三方评估促进生态治理现代化

当前，我国正处于大力推进生态文明建设的关键时期。实现生态文明领域国家治理体系和治理能力现代化，必须客观评估各地生态文明发展水平，总结成功经验，寻找关键制约因素，从而有效推进生态文明建设。评估生态治理的效果和水平，不仅要关注生态环境领域本身，还要统筹生态环境与经济、政治、文化、社会等多个领域的关系，建立科学的评价指标体系。生态文明领域的治理体系和治理能力是评估中两个尤为重要而不可分割的部分，治理体系的完善是治理能力提高的基础；治理能力及实践能力的提高是治理体系发挥效用的关键。

生态治理体系评估

评估生态治理体系的完善程度，需要统筹经济、政治、文化、社会及生态建设各个领域，综合考量生态治理的各个环节及参与治理的各个主体。因此，其评估体系不仅需要具有长期遵守的原则标准，还需要动态调整的灵活机制，因地制宜、因时而变。

首先，须系统评估生态治理制度体系的完备程度。尽管我们已经在绿色发展、生态环境保护方面制定了诸多相关制度，但其系统性、完整性尚待完善：源头上缺乏有效防范制度，自然资源资产的产权制度尚未完全建立；过程中缺乏严密监管制度，污染物排放许可制亟待健全；结果上缺乏严厉的责任追究和赔偿制度。生态文明制度是约束人类行为的规则，同时也是衡量人类文明水平的标尺。治理环境的高额度投入固然能使生态环境得到一定程度的改善，但如果环境法律法规标准、公众的生态环境意识停滞不前，那么生态环境治理只能治标不治本、生态文明水平只是浮华虚高。良好的生态环境是生态文明的硬实力，先进的制度体系则是生态文明的软实力。制度体系是否系统和完整，是否具有先进性，在一定程度上代表了生态治理现代化水平的高低。

其次，生态治理的标准化建设程度评估是生态治理体系评估的重中之重。国家治理现代化要求标准化、制度化、法治化，标准化是其中的重要环节。同样，在生态治理领域，只有具备了统一且行之有效的标准体系，生态治理决策和管理过程才有规可依。从生态目标的设定、治理内容的具体化到治理效果的评估，每个环节都需要标准化的流程，需要建立可依据、可量化、可执行的指标体系。指标体系作为一个参照系统，其结构和组成要素的组合方式会直接影响系统功能的发挥。因此，它必须涉及生态空间、生态经济、生态环境、生态生活、生态文化、生态制度等各个方面，涵盖能耗、水耗、地耗、污染物排放、环境质量等标准，具有层次清晰、可定量评价、可获取性强等特征。

生态治理能力评估

生态治理能力是生态治理体系在具体实践中的体现，生态治理能力评估可以分为结果评估、过程评估和主体评估三个层面。

第一，治理结果评估。以往的治理结果评估，一般仅包括工业废水排放达标率、生活垃圾无害化处理率等具体的环境和经济指标，而常常忽视了生态治理的目标应是"人与自然"的和谐共存、协同发展。在治理结果评估中，应重视公众满意度这一评价指标，重点考察公众的生态环境满意度、污染治理满意度和政府生态治理效率满意度。同时，要丰富公众的参与方式，充分结合传统问卷调查与网络平台调查，调查过程中兼顾被调查群体的全面性，力求最大限度地反映不同利益群体的诉求。

第二，治理过程评估。对于生态治理能力的研究，长期流于治理结果层面，对于治理过程则关注不足。但生态治理是一个动态的系统发展过程，并没有静态的均衡结果，过程与结果在动态系统中是统一的。生态治理过程可以分为两大类，一是对既有资源开采过度和环境污染问题的治理；二是对生产和生活消费过程中新增生态问题的治理。前者可视为存量治理，后者可相应地视为增量治理。

存量治理的过程评估相对简单，评价指标包括生态治理资金投入、生态治理决策的民主程度等。资金投入是生态治理的物质基础，一切生态治理工程的启动与实施都依赖于充足资金的支持。而决策的民主程度则是生态治理的科学保障。民主程度可以通过相关领导干部选拔任用是否经过民主推荐和公开考察，决策的形成是否经过利益相关主体的多方磋商以及信息是否公开透明来体现。民主程度的高低，关乎治理过程的程序正义，也是提升政府主导下多元主体生态治理模式公信力的关键。

而增量治理的过程评估，除了需要进行存量治理评估之外，还需要走进生产生活的一线进行考察评估，全面了解现实存在的新增生态问题，把握其现状及治理力度。对生产过程治理进行评估，可以在重点企业中引入产品生命周期评价和物质流分析等方法，从而清楚地了解各个生产环节的投入产出状况及其环境影响；对生活消费过程治理进行评估，则应以社区为单位，全面地评估垃圾分类回收、建设节能社区等生态建设举措的贯彻落实情况，从而全面地掌握基层社区的生态文明现状以及民众的生态诉求。

第三，治理主体评估。传统的生态治理模式为政府自上而下的管制模式，不可避免地会带来治理决策失误、执行不力、监管不足等问题。同时，权力寻租现象屡禁不止，造成政府与企业、民众之间利益分化，进而导致政府公信力的下降。生态环境具有公共产品属性，生态治理也是关乎国计民生的重大议题，因此，生态治理的过程中需要有居民、企业、非政府组织、媒体等多元利益主体的广泛和深入参与。

不同的治理主体，治理能力要求各不相同。要根据不同的治理主体构建差别化治理评估体系。治理能力现代化评估最重要的方面应该是对党和政府这一治理主体的治理能力的评估。在国家治理体系中，执政党与政府是最重要的治理主体，对党和政府治理能力现代化水平的评估主要是制度设计能力、监管能力、危机应对能力等。同时，其他治理主体的作用和影响也不容忽视，应逐渐规范化纳入评估体系。

引入第三方评估机制

第三方评估,即由独立于政府及其部门之外的第三方对生态治理体系和治理能力进行评估。作为一种必要而有效的外部制衡机制,第三方评估由于较好地体现了其独立性与专业性的优势,避免政府部门既当"运动员"又当"裁判员"的弊端,弥补了政府自我评估的缺陷,所得出的科学客观的判断结论更有可信度、说服力,可以增强社会监督力量,提升政府的公信力。

必须加强第三方评估的制度保障。

一是坚持评估机构的独立性。评估主体的中立角色是保证结果公正的最基本前提。评估主体如与公共权力融合紧密,则政府难避作秀诟病。应制定第三方评估单位遴选办法和对第三方机构的监督管理办法,定期公布第三方评估机构名单,让第三方机构竞争公平、工作透明,从而提高公信力、可靠性。

二是确保评估的专业性。第三方评估机构不仅具备独立的特性,同时也应是生态文明建设领域的专业机构。构建评估指标体系、确立评估标准以及确定评估实施方案,每个评估的关键环节都具有很高的专业技术含量。评估指标和方法、数据来源、结果必须具有可检验性,对评估信息事前公示,对评估结果事后公布,公开接受社会各界的品评和检验,建立开放和接受监督的评估系统。

三是要使评估结果能得到有效的运用,决不能让第三方评估报告"束之高阁",要与问责机制、干部考核等直接挂钩、密切结合,最大限度地解决职责不清、推诿扯皮、工作不落实等问题。对于评估中发现的问题,专业的评估机构可给予精准导向,提出"对症下药"的解决方案,更加准确明确、清晰有度地落实部门责任。

Chapter 6
Utilize Third-party Evaluation to Promote Ecological Governance Modernization

Currently, China is in a critical period of vigorously promoting ecological civilization construction. To modernize the state governance system of and ability in ecological civilization, it is necessary to objectively evaluate the development level of ecological civilization, draw on successful experience and find the key restrictions, so as to effectively promote the ecological process. To evaluate the effect and level of ecological governance, we should not only pay attention to ecological environment, but also associate it with economic, political, cultural, social and other aspects to establish a proper evaluation system with scientific indicators. The governance system and governance capability are two important and inseparable parts of the evaluation of ecological civilization: the perfection of governance system is the foundation for improving governance capability; the improvement of governance capability and practical ability is the key to give play to the governance system.

Evaluating the Ecological Governance System

The evaluation of ecological governance system integrates economical, political, cultural, social, ecological construction and other fields, and comprehensively considers various links and subjects involved. Therefore, its evaluation requires not only long-term principles and standards to follow, but also a dynamic and flexible mechanism to allow adjustments according to local conditions, times and changes.

First, the readiness of systems concerning ecological governance should be evaluated. Although a number of systems in green development and ecological environmental protection have been adopted, they have yet to be improved more systematically and comprehensively: there is a lack of effective preventive system on the top and an incomplete property rights system of natural resources; there is a lack of strict supervision in the process and underdeveloped pollutant discharge permit system; there is a lack of system for strict accountability and compensation at the end. A system for ecological civilization comprises human behavioral rules and is the gauge of human civilization. High amount of investment could improve ecological environment to some degree. However, if environmental laws and regulations and public awareness towards ecological environment remain stagnant, ecological governance will be a palliative and high-level ecological civilization a bonfire of vanities. A good ecological environment is the hard power of ecological civilization while an advanced system is its soft power. The extent of systematic readiness and advancement of ecological governance determine its level of modernization.

Second, the assessment of the degree of standardization of ecological governance is the top priority during the evaluation of ecological governance system. Modernization of national governance requires standardization, institutionalization, and rule of law, among which standardization is one of the key parts. The same applies to ecological governance. Only by setting up a group of unified and effective standards, can there be rules to abide by when it comes to decision-making and management during the process. From the setting of ecological goals, the specification of content, to the assessment of effects, standardized processes and a reliable, quantitative, executable system of indicators are needed. Indicator system is a reference system whose structure and combination of elements will directly affect its function. Therefore, the indicator system must involve ecological space, ecological economy, ecological environment, ecological life, ecological culture, ecological policies and other aspects, covering energy consumption, water consumption, land consumption, pollutant emissions, environmental quality and other standards. Moreover, the indicator system should be distinctively classified, quantitative and accessible.

Evaluating the Ecological Governance Capabilities

The ecological governance capability is reflected in the practices of ecological governance system. The evaluation of ecological governance capability includes three parts: result evaluation, process evaluation and subject evaluation.

First, the governance result evaluation. In the past, the evaluation of governance results generally includes only environmental and economic indicators such as industrial wastewater discharge rate and household garbage biosafety disposal rate. What has been overlooked is that the goal of ecological governance is to achieve harmonious coexistence and synergic development between "human and nature". During the evaluation of governance results, we should highlight the indicator of public satisfaction, focusing on public satisfaction on ecological environment, pollution control and government efficiency. At the same time, the way to involve public participation should be enriched, combining traditional questionnaire and online platform. The participants into research should be as inclusive as possible to demonstrate different interests of different groups as much as possible.

Second, the governance process evaluation. The study of ecological governance capability has long focused on the results and neglected the governance process. Since ecological governance is a dynamic and systematic without static equilibrium outcomes, the process and results are dynamically unified. Ecological governance process can be divided into two categories: one is the management of over-exploitation of existing resources and pollution problems; the other is the management of emerging problems during production and consumption. The former can be regarded as stock governance and the latter, increment governance.

The process evaluation of stock governance is relatively simple. Its indicators include the investment into ecological governance, the degree of democracy in decision-making and so on. Investment is the material basis for all projects of ecological governance to start and implement and democracy in decision-making offers scientific protection for ecological governance. The degree of democracy embodies in democratic nomination and public inspection of related leaders, consultation with multiple stakeholders in decision-making, and information transparency. The level of democracy is closely related to procedural justice, and is the key to enhance the credibility of multi-subject ecological governance led by government.

The process evaluation of incremental governance involves not only the elements of stock governance, but also field inspection during production and living to get a comprehensive understanding of emerging ecological problems, their current status and managing intensity. During the evaluation of production process, product life cycle assessment and material flow analysis can be adopted for key enterprises to clearly understand the input-output and environmental impact of various links in the production chain. Meanwhile, the evaluation of living and consumption processes should contain a comprehensive assessment of waste separation and recycling, energy conservation and other ecological practices by communities, so as to comprehensively grasp the ecological civilization and public demands of grassroots communities.

Third, evaluation of the governing bodies. Previous ecological governance followed a top-down mode managed by the government, inevitably causing decision-making mistakes, poor implementation, inadequate supervision and other issues. At the same time, rent-seeking was rampage, resulting in separation among the interests of government, businesses and the public, which in return led to the decline in government credibility. Since ecological environment features public goods and ecological governance is related to the national economy and the people's livelihood, the process of ecological governance need wide and in-depth participation of residents, companies, non-governmental organizations, media and other stakeholders.

Requirements of governance capabilities vary for different governing bodies whose evaluation system should also be differentiated. The most important aspect of the evaluation of modern governance is the assessment of governance capabilities of the Party and the government. In a governance system controlled by the state, the ruling Party and the government are the most significant governing subjects. The evaluation of the modern governance capabilities of the Party and the government includes the assessment of their system design ability, supervision ability, and the emergency handling ability. Meanwhile, the role and influence of other governing bodies cannot be ignored. Instead, they should be standardized and incorporated into the evaluation system.

Introducing Third-party Evaluation Mechanisms

Third-party evaluation draws on agencies independent of the government and official authorities to assess the ecological governance system and governance capabilities. As a necessary and effective external mechanism for checks and balances, third-party evaluation has independence and professional advantages to avoid drawbacks of governmental departments who act both as "athletes" and "referees". It makes up for the defects of government self-assessment with scientific, objective, reliable, and convincing conclusions and judgments which can enhance the power of social supervision as well as the credibility of the government.

Institutional assurance for third-party evaluation must be strengthened.

First, independence of the third parties must be maintained. The neutrality of evaluator is the basic prerequisite to ensure the fairness of the results. If the evaluator is related to public power closely, government will be criticized for attitudinizing. Therefore, rules to select, supervise, and manage the third parties should be made. The list of selected third-party agencies should be released regularly, which could ensure fair competitions among them and transparent operation, so as to improve the credibility and reliability of the evaluation.

Second, the third parties must be professional. Third-party evaluation agencies not only have to be independent, but also be specialized in ecological civilization construction. The establishment of indicators, criteria and implementation plan, and other key steps of evaluation are professionally demanding. Indicators and methods, data sources, and results must be verifiable. The information to be evaluated should be publicized in advance and the results of evaluation are publicized afterwards, open to public review and scrutiny, to ensure an open and evaluation system under supervision.

Third, the evaluation results and reports should be effectively used, instead of being cast aside. They should be closely tied to accountability mechanism and cadre appraisal system, to solve issues such as unclear responsibilities and duties, buck-passing, and inaction to the greatest extent. For problems found during the evaluation, the third parties may give precise guidance and corresponding solutions where departmental responsibilities and duties are specified and assigned.

第七章　构建多元协同的生态治理模式

推进生态治理现代化是一项复杂的系统工程，涉及社会的各个领域，关乎参与其中的各个角色。由于生态环境具有整体性，这就要求改变传统的政府单一治理模式，积极利用市场主体及培育各类社会组织，鼓励非政府力量参与到生态共治之中，最大限度地维护和增进公民利益，形成政府、企业、社会组织和公众等多元化权利主体治理体系。政府要更加明确自身的职责和权力范围，从全能政府转化为有限政府，将治理权利和工具分配到公众、社会组织等主体中。

随着人们生态环保意识的不断增强，生态建设人人有责的理念已经在全社会形成共识。然而，由于实践中存在法律制度滞后、政府职能转变不到位、信息不对称、各主体利益诉求不同等多种因素，生态多元共治往往失效。新形势下，探索多元协同的生态治理模式既是形势所迫，也是大势所趋。

生态治理需要多元主体参与协同

"良好的生态环境是最公平的公共产品，是最普惠的民生福祉。"生态环境作为公共产品，具有效用的不可分割性、受益的非排他性和消费的非竞争性。一方面，非排他性使得"搭便车"问题难以解决，非竞争性又给公共产品定价带来困难，仅仅依靠市场力量难以保障良好生态环境的供给；另一方面，政府单一供给公共产品不仅缺乏竞争机制，而且还会导致不同程度的寻租行为，具有一定的局限性。因此，政府与社会力量在供给良好生态环境的过程中不是非此即彼的关系，而是协同共治的良性互动。

广义上讲，生态治理是为了克服资源无度享用、经济粗放发展、环境污染破坏的制度安排，其本身也是一种公共事务。由于生态问题的复杂性，在制度安排上要考虑空间、时间、主体等多个方面，需要因地制宜、因时制宜，以满足不同主体的生态环境需求。只有逐步完善生态共治的政策与制度体系，才能够建立起良性、协同、可持续的生态新秩序。可见，解决生态环境保护及生态环境制度两大问题的关键点，就在于协同多元主体，形成激励机制，催生目标一致的高效行动力。

生态共治多元主体的角色担当

一要发挥政府在生态治理中的关键作用。生态问题的公共性决定了政府必然要在生态文明建设中起到关键作用，发挥协调、组织、管理、推动的职能，制定多元治理体系的宏观框架和参与者的行为准则，运用经济、法律、行政等多种手段为公共事务的处理提供依据。尽管政府需要从"控制型"向"服务型"转变，其在公共事务管理中的话语权和控制力仍然具有其他主体所不能替代的优势。要在生态治理中发挥横向总揽全局、协调各方的作用，同时要处理纵向上中央与地方的关系，打破"上有政策，下有对策"的局面，走出"一放就乱，一乱就收，一收就死，一死就放"的循环。

二要调动企业和各种市场主体的积极性。企业作为"工业时代"的行动主体，创造人类发展辉煌的同时也成为"谋杀生态的元凶"。然而，由于其市场导向的营利性天然目标，企业具有最强、最快、最积极的行动力。因此，应积极推动生态环境影响等外部性问题内部化，引入竞争机制，创新融资方法，使合作参与者提前全面考量成本利益。当一个成熟的企业深入认识到生态环境与其长期利益追寻的关系，就会自然而然地做出牺牲短期利益实现生态保护的抉择，主动与社会各界共同承担建设生态文明的重任。政府须探索利用市场化机制推进生态环境保护，推广政府和社会资本合作模式，引导企业生态友好化的逐利行为。

三要大力发挥社会组织的价值。如果说国家代表了公共领域管理者，公民代表了私人领域所有者的话，那么，社会组织就是沟通公共领域与私人领域的"纽带空间"，是独立于国家和公民之外的"第三空间"。提倡和培育以非政府组织（Non-Governmental Organizations，NGOs）、非营利组织（Non-

Profit Organizations，NPOs）为代表的公益性社会组织，是生态治理的必然要求。从国际经验来看，欧美国家 NGO、NPO 数量众多、规模不等、宗旨各异，通过其自身优势，在生态环境保护中发挥着不可替代的作用，成为政府力量的有效补充。大型组织如大自然保护协会（The Nature Conservancy，TNC）是国际上最大的非营利性的自然环境保护组织之一，管护着全球超过50万平方公里的1600多个自然保护区，8000公里长的河流以及100多个海洋保护区。自1998年进入中国以来，TNC 和中国政府广泛合作，引入国家公园概念，进行国家公园探索示范；吸引社会资金进行碳汇造林，缓解和应对气候变化。这类活跃在我国的专业社会组织在引入先进国际理念和方法、创新保护模式、吸引和带动民间资本投入环境保护、提升国内社会组织的能力水平方面发挥了积极的作用。应该发挥市场机制，通过政府购买公共服务等多种方式鼓励社会组织主动参与生态环境治理，形成多元化的投资模式和治理主体。

四要充分调动公众的力量。生态环境与每一个人的生存和发展密切相关，个体既是生态危机最直接的受害人，也是生态治理中最广泛、最基础的实践主体和力量源泉。只有获得广泛而有效的公众支持，生态文明建设才能取得广泛而持久的成果。作为生态环境利益的直接相关者，个人可以对企业的生态破坏行为进行直接监督，或向政府提出诉求，进行间接监督，更可以通过多元利益合作组织平台，将零星的个体力量整合起来，有效防治生态破坏行为。完善公众参与机制，提高专业人才的话语权和影响力，对保障公众对生态治理的知情权和参与权具有重大的现实意义。

完善协同共治的运行机制

推进生态治理现代化，使生态治理的多元主体真正发挥相应的作用，产生 1+1>2 的效果，避免陷入各自为政和相互掣肘的局面，就必须要有一套完善的协调机制。

一要保障各主体的自治权，夯实生态协同共治的基础。主体的自治权是共治的基础。如果没有主体的自治，共治是难以实现的，因此应当遵循主体自治与共治相结合的原则。我国历来有大政府的传统，政府必须要尊重其他主体的权利，为多元共治的实现提供一个良好的平台。社会组织的发展，在符合法律规范的基础上，政府不应该过多干预，更不应该随意取缔；企业作为重要的市场主体，政府可以适当引导，但不能代替决策；公民是最广泛的治理力量，政府应当切实保障其知情权，引导公民参与的有序进行。

二要统筹各主体权责，确立生态共治机制。生态共治机制的核心是明确各主体的权力（权利）与责任，从而保证共治机制得以正常运转。就整个系统而言，它必须是互动融合的开放性复杂系统，必须具有灵活性和多样化的特征。从权力（权利）行使来看，要做到让各个主体"在其位，谋其政"，理顺中央政府和地方政府之间、政府各部门之间、政府和企业之间的关系。为保证各主体在整个生态治理中的参与度，就需要将它们纳入到综合决策机制中来，不仅生态治理议题的提出、信息的披露与共享，而且决策的过程都要充分考虑各主体的意愿，尊重每一个主体在整个治理过程中的利益诉求。从责任分配来看，要将生态治理责任按权力（权利）合理分配，确保各主体承担相应的责任，同时也要避免"各人自扫门前雪，不管他人瓦上霜"的情况。

三要加强责任监管，维护生态共治体系。责任监管首先要推进生态环境及治理信息公开。在生态文明建设领域，信息公开的关键主体是政府，其透明的程度可以说是协同治理能否实现的关键性因素。在信息化、现代化过程中，多数国家选择以立法的方式促进环境信息公开。在此基础上，还应通过行政、司法、媒体等渠道，利用新媒体等宣传载体，畅通信息传输渠道，对参与共治的各主体予以监督和有效管理；通过完善司法监督与社会监督机制，严格落实生态惩戒举措，保证各主体能够履行其生态治理责任。

Chapter 7
Establish a Model of Ecological Governance: Coordinating All Stakeholders

The modernization of ecological governance is a complex and systematic project that involves roles from all walks of life. The integration of ecological environment requires a change in traditional governance model managed only by the government and actively draws market players, different kinds of social organizations, and non-governmental institutions into ecological co-governance, to maximize public interests. By doing so, a coordinated ecological governance system could be set up, comprising government, enterprises, social organizations, public and other subjects of the right. The government should clarify its responsibilities and powers, converting itself from an omnipotent government into a limited government and distributing power and tools to the public, social organizations and other subjects.

With growing awareness of environmental protection, the concept that ecological construction is of everyone's responsibility has been well-accepted by the whole society. However, due to the limitations such as a lagged-behind legal system, insufficient transformation of government functions, information asymmetry, and interest conflicts among different subjects, ecological co-governance is often ineffective. Under the new situation, it is objectively and subjectively necessary to explore a model of multi-subject coordinated ecological governance.

Ecological Governance Needs Multi-stakeholder Participation

"A wonderful ecological environment is the fairest public goods and the most inclusive welfare to people". As a public product, the utility of ecological environment is indivisible; its benefits are non-exclusive; and its consumption is non-competitive. On one hand, non-exclusiveness leaves a stubborn loophole for "free riders" and non-competitiveness makes its pricing difficult. The power of market is not enough to guarantee the supply of good ecological environment. On the other hand, if government is the only supplier, there will be some limitations, such as a lack of competition mechanism and rent-seeking. Therefore, government and social forces are not mutually exclusive but beneficially interactive in providing a good ecological environment.

Broadly speaking, ecological governance is the institutional arrangement to overcome overexploitation of resources, extensive pattern of economic development, and environmental pollution, which proves itself to be inherently a public affair. Due to the intricacy of ecological problems, it is necessary to consider the location, time, subject and other aspects. That is to do what is suitable to the location and to the particular time and meet different needs from different subjects. Only by gradually improving the policy and system of ecological co-governance can a new, benign, synergic, and sustainable ecological order be built. It's fair to say that the key to manage ecological environment protection and ecological environment system is to coordinate multi-subject participation and adopt proper incentives to encourage efficient execution under the same purpose.

Roles of Different Stakeholders Under Co-governance

First, the government plays a key role in ecological governance. The public nature of ecological problems determines government's leading role in ecological civilization construction, including coordination, organization, management and promotion. It is responsible for formulating a macro framework for ecological co-governance and a code of conduct for participants and taking economic, administrative and other means to support the handling of public affairs. Although the government needs to transform from a "controlling" role to a "service-oriented" one, its authority and control force in public affairs are dominating and irreplaceable by other subjects. In ecological governance, the government has to horizontally control the overall situation and coordinate different participants, and vertically deal with the relationship between the central and local authorities. "Where there is a government policy, there is a way to get around it" is not allowed. The vicious cycle of "Decentralization – Disorder – Centralization – Standstill – Decentralization" must be broken.

Second, enterprises and various market players must be motivated. Enterprises are the main actors in the "industrial era". They create a brilliant human development while being the "culprit of ecological degradation". Due to their market-driven and profit-oriented natural, enterprises have the strongest, fastest and most active motivation. Therefore, we should internalize the external issues of environment, by introducing competition mechanism and innovating financing methods to allow participants to consider cost effectiveness in advance. When a mature enterprise gains an in-depth understanding of the relation between its long-term benefits and ecological environment, it will compromise short-term interests over ecological protection and shoulder its responsibility along with other sectors to build ecological civilization. Meanwhile, the government needs to employ the market-based mechanism to promote ecological protection and encourage the cooperation between official and social capital, guiding the eco-friendly profit-making of enterprises.

Third, the value of social organization should be explored. While the state dominates the public sector and citizens represent the private sector, social organizations are the bridge connecting the two sectors and "the third space" outside the state and citizens. To advocate and nurture Non-Governmental Organizations(NGOs), Non-Profit Organizations(NPOs), and other social organizations for the public good, is the inevitable requirement of ecological governance. Looking abroad, there are numerous NGOs and NPOs with different sizes and objectives. By enjoying their advantages, they play an irreplaceable role in ecological protection as an effective supplement to government power. For example, The Nature Conservancy (TNC), one of the world's largest non-profit organizations for nature conservancy, is protecting more than 1600 natural reserves of over 500000 square kilometers, rivers of 8000 kilometers, and over 100 marine reserves. Since its first presence in China in 1998, TNC and Chinese government have conducted extensive cooperation in introducing and building national parks and in attracting social funds to carry out carbon sinks and relieve climate change. Such kind of social organizations with presence in China has played an active role in introducing advanced ideas and methods, innovating the model of protection, driving private capital into environmental protection, and enhancing the ability of domestic social organizations. Therefore, the government should capitalize on market mechanism to encourage social organizations to take part in ecological governance through multiple ways such as buying public services, so as to form a diversified investment model and creating multiple governing subjects.

Fourth, the public should be mobilized. Ecological environment is closely linked to the survival and development of each individual. Individual is the victim most directly affected by ecological crisis and the most extensive and basic implementer and source of power in ecological governance. Ecological governance will not achieve extensive and lasting outcomes without wide and effective public support. As a stakeholder of ecological environment, individuals can directly supervise ecological damages caused by enterprises or indirectly supervise by resorting to the government. They can also take advantage of cooperative platforms for groups with multiple interests, integrating individual forces together to effectively redress ecological damaging activities. In conclusion, public participation mechanism should be improved. Voices from professionals and experts should be heard and their influence be expanded. It bears great realistic significance to ensure public's right to know and right to participate during ecological governance.

Improving the Operating Mechanism of Co-governance

An improved coordinating mechanism of co-governance is necessary to promote the modernization of ecological governance, give full play to multiple governing subjects, generate an expansion effect of $1 + 1 > 2$, and avoid becoming fragmented and mutually restrictive.

First, the autonomy of the each stakeholder should be guaranteed to consolidate the basis of ecological co-governance. The autonomy of each player is the basis of co-governance. If there is no autonomy for subjects, co-governance is difficult to sustain. Therefore, autonomy of player and co-governance among them should be combined. To transform the tradition of "big government", Chinese government shall respect the rights of other subjects to provide a good platform for multi-player co-governance. The government should not intervene too much in or arbitrarily forbid the development of social organizations which maintain legal compliance; the government can properly guide enterprises which serve as important market players instead of making decisions for them; since citizens possess the most extensive power of governance, the government should effectively protect their right to know and involve their participation in an orderly manner.

Second, rights and responsibilities of each stakeholder should be clarified to support the ecological co-governance mechanism. The core of ecological co-governance mechanism is to clarify the power (rights) and responsibilities of the stakeholders so as to ensure the normal operation of co-governance. The whole system must be interactive, complex but open, flexible and diverse. For the execution of power (rights), it is necessary to make each stakeholder "do what should do". The relationships between the central government and local governments, among governmental departments, and between government and enterprises should be straightened out. In order to ensure the participation of different players in ecological governance, these players need to be integrated into the comprehensive decision-making mechanism, not only in proposing ecological agenda, the disclosure and sharing of information, but also in decision-making process by respecting the interests of each stakeholder in the whole process. From the perspective of responsibility allocation, it is necessary to allocate the responsibilities of ecological governance according to power (rights) to ensure that each governing subject bears corresponding responsibility and to avoid the indifference that "everyone only takes care of his/her own business".

Third, supervision should be strengthened to uphold the ecological governance system. To achieve this, the first thing is to promote the disclosure information about ecological environment and governance. Government is the main body for disclosing information, and the degree of transparency is the key in realizing co-governance. In an era of information and modernization, most countries choose to promote the disclosure of environmental information through legislation. Apart from doing the same, we should also supervise and effectively manage various players through administrative, judicial and media channels and by using new media and other information carriers to smoothen information spreading. In addition, we should ensure that each player fulfills its ecological responsibility through the improvement of judicial supervision and social supervision and by strictly implementing disciplinary action against wrong-doing.

第八章 地方政府竞争转向助推生态治理

地方政府竞争模式是改革开放以来形成的极具中国特色的发展模式。各地方政府通过土地、税收优惠等手段，竞相招商引资，极大地提升了经济发展的速度，并使基础设施建设不断完善，但政绩考核机制导致地方政府的"逐底竞争"行为，使得生态环境长期超负荷运行，恶化迅速而且前景堪忧。在生态—经济协同发展的新思路下，应充分利用地方政府竞争模式的优势，通过一系列激励相容的制度安排，挖掘地方政府在生态环境治理方面的潜力，诱导形成"分权治理＋地方竞争"的中国特色生态环境治理模式。

何谓地方政府竞争

1994年分税制改革以后，地方政府独立预算、自收自支、自求平衡，逐渐实现了中央政府与地方政府之间税种、税权、税管的划分，实行了财政"分灶吃饭"。在"地方税"及"中央—地方共享税"的激励机制下，地方政府具有强烈的意愿提升本地吸引力，通过发展本地经济、扩大税基来增加税收收入，进而提升地方政府提供公共产品的能力，从而依此逻辑形成本地发展的良性循环。纵观这一逻辑链，关键起点为发展本地经济，行之有效的方法就是积极提供更完善的基础设施、更高比例的财税优惠和更高效率的行政审批，以此增强地方竞争力，吸引国内外资本、劳动力、技术等生产要素，推动本地经济发展。同时，由于中央政府仍掌握着包括政策、资源、税收等方面大量的优惠权，地方政府如果能向中央政府争取到某些优惠政策和特殊待遇，就能为地方经济的发展创造重要机遇。因此，谋取中央政府提供的优惠政策、争取有利的政策环境成为地方政府竞争的重要内容。

分权制度与地方政府竞争模式从设立至今为我国经济的高速发展做出了重大贡献，各地得以迅速建立起相对完善的基础设施体系，政府部门工作效率大幅提升，居民收入水平显著提高。这一模式还充分发挥了地方政府的信息优势，小型的改革试验在各地如火如荼地展开，为全国范围的深化改革创造了巨大的创新空间。但是地方政府竞争也给生态环境带来一系列负面影响，地方政府或忽视环境保护，或过度利用自然资源和生态环境，甚至放纵高污染企业破坏环境。

地方政府竞争为何在过去带来了严重的生态问题

在促进我国经济快速发展的同时，各地方政府竞争也因为对资本的过度追逐而导致普遍出现只注重经济利益，甚至是"杀鸡取卵""竭泽而渔"，忽视生态环境保护。为什么过去三十多年我国的地方竞争会给生态环境带来如此严重的污染和破坏呢？直观上来看，是由于地方政府竞争体系的不完善。地方政府竞争以政绩考核为标准，而以往的地方政绩考核未将生态成本、生态建设成果纳入其中。同时地方竞争缺乏有力的约束机制，在行政决策体制、决策程序、责任追究制度等方面缺乏法律制约。在这样的竞争体系下，一些地方政府追逐经济利益和社会利益，缺乏治理生态环境的动力。

从更深层次而言，地方政府缺乏治理生态环境的动力还与地方政府竞争条件下生态治理的外溢效应和"污染避难所"效应密切相关。生态治理的外溢效应是由生态治理的本质所决定的，即将生态环境作为公共物品向公众提供的过程。地方政府通过执行严格的环保标准、加大违法排污的处罚力度、投资建设环保设施等方式治理本地环境，就是为老百姓提供了人人可以享用的公共物品——良好的生态环境。然而，由于公共物品具有消费的非排他性和非竞争性，其提供常常伴随着"搭便车"行为，生态治理也不例外。本地的生态治理成果跨越了地区之间的行政分界而外溢，使得邻近地区也能免费享受到清洁的蓝天和河流所带来的好处，这种现象就是生态治理的外溢效应。为了在地方政府竞争中胜出，地方政府坐等"邻居"提供生态治理"产品"而自身怠于行动是最"理性"的，而这无疑会导致"三个和尚没水喝"的情形。

与此同时，在地方政府对资本的恶性竞争中，生态治理无疑会削弱本地的相对竞争力，从而产生"污染避难所"效应。如果本地执行更加严格的环保标准，便极有可能导致污染企业迁移到附近环保标准相对较低的地区，且由于迁移地区环保标准更低，企业可以更肆无忌惮地排放污染。那么站在区域和全国的角度看，单独一个地市的严格环境政策反而加重了污染总量。地方政府治理生态环境不仅需要动力，也需要能力。而当前中央与地方的生态环境事权与财权不匹配，也会导致生态环境治理效果欠佳。

如何诱导地方竞争保证生态治理

在地方竞争模式下，地方政府缺乏治理生态环境的动力和能力。尤其是由于外溢效应和"污染避难所"效应的存在，在生态环境治理过程中"自掏腰包"往往会损害本地的经济发展前景，甚至反而加剧全国的污染，因此对于地方政府和官员而言并非一个明智的决策。然而这一模式必然会导致生态环境破坏吗？恐怕也不尽然，如果可以扭转外溢效应和"污染避难所"效应，恰当地通过一系列激励相容的制度安排，调整地方政府治理环境的动机，同时提升地方政府治理环境的能力，就能诱导其竞相提供更好的生态环境。

首先，要提高地方政府治理环境的收益支出比，促使生态治理成果成为地方政府的竞争优势。要将对地方政府治理环境成效的评价与该地投资环境、上级财政转移支付、地方政府债务贷款等因素挂钩。在投资环境方面，将生态环境指标纳入投资环境评价体系，引导投资者形成关注生态环境的正确态度，促使形成生态环境对本地获取财政和资本的支持效应，从而将生态环境提升到与基础设施建设、政府办事效率等地方政府竞争重点同等的位置上。在财政转移支付方面，明确划分中央政府与地方政府之间的生态环境事权财权，规定区域内的生态环境问题由地方治理，跨区域的问题由中央统筹各有关地方政府进行治理，保证事权与财权相对应。

其次，要改革政绩考核制度和竞争约束机制，推进地方政府竞争向生态转向。以往以 GDP 为主的地方政府考核体系主要集中在生产总值增长率、税收上缴量、吸引外商投资额等指标上，使得地方政府和官员对考核体系外的事务较少主动关心，加上决策体制、决策程序等不完善，使得地方政府和官员的意志成为地区经济社会发展的决定性力量，所以发展地方经济的同时常常以牺牲环境质量为代价。要进一步改革政绩考核指挥棒，将公众和环保组织的意见纳入考核范围，增加生态环境类问题的信访数量、环保社会组织发展状况、环境信息公开程度等硬性指标，同时建立健全地方政府生态治理决策体制、决策程序、责任追究制度等方面的法律法规，激励地方官员更加重视本地生态环境治理工作。

最后，要在相关区域内执行统一的环保标准来避免污染避难所效应。建立区域联防联控制度，统筹协调区域内各行政区的具体情况，采取统一的环保标准，创新地方政府间的区域性污染物排放权交易制度。建立地方政府间环境治理互评监督制度，抑制和防止恶意降低本地环保标准执行力度来吸引投资的行为，诱导地方政府间生态环境治理的竞争从"趋劣竞争"向"趋优竞争"转化。推进主体功能区建设，根据区域核心功能构建环境规制与政绩考核制度，同时考虑不同区域的生态环境承载力，因地制宜地制定区域发展规划和环境保护政策。

我国已进入工业化后期，按照经济发展规律和国际经验，在我国经济结构性改革和产业升级的大背景下，绿色发展将逐步成为经济发展的新动力。随着基础设施的不断完善和政策可优惠空间的缩小，各级地方政府间的竞争焦点将逐步转向技术创新、改善生态环境质量等方面。如果能通过制度规定来影响竞争规则，积极诱导改变地方竞争的发展方向，将会极大地挖掘地方生态治理潜能，提升生态环境治理水平。

Chapter 8
Transform from Local Government Competition to Ecological Governance

Encouraging local government competition is a development path with strong Chinese characteristics since China's reform and opening up. Local governments, through land and tax incentives, compete for foreign and domestic investment, which accelerate economic growth and improve infrastructure construction. This path and the political performance appraisal mechanism, however, created a race-to-the-bottom effect, driving rapid deterioration of the ecological environment. Under the new thinking of coordinated development between ecology and economy, the advantages of local government competition should be fully taken. A series of encouraging and compatible institutional arrangements should be introduced to tap the potential of local governments in ecological governance. The aim is to build an ecological governance model featuring "decentralization and local competition".

What Is Local Government Competition?

After the reform of the tax system in 1994, local governments exercise independent budget planning and take responsibility for their own financial revenues and expenditures to achieve fiscal balance. Gradually, a tax assignment system on the basis of a rational division of power between central and local authorities has been built. Under the policy of "local tax" and "central-local fiscal sharing", local governments have a strong desire to strengthen local competitiveness, increase tax revenue by developing local economy and expanding tax base, enhance their ability to provide public service and create a virtuous circle for local development. This logical chain starts by developing local economy through effective ways, such as providing better infrastructure, higher proportion of tax incentives and more efficient administrative approval procedures, in order to enhance local competitiveness, attract such factors of production as foreign and domestic capital, labor force and technology. Since the central government holds a large amount of preferential policies on resources and taxation that are crucial for local economic growth, local governments compete fiercely for such policies.

The decentralization system and local government competition have made great contributions to the rapid development of China's economy since its establishment. Local governments have shortly built a relatively complete infrastructure network, improved government efficiency and increased the resident's income. This model also allows local governments to give full play to their advantages in gathering information. Small pilot reforms are carried out in full swing, which creates room for innovation in deepened reforms. The local government competition, however, also brings a series of negative effects to the ecological environment. For instance, local governments may ignore environmental protection, overuse natural resources, or even tolerate environmental damage by high-polluting enterprises.

Why Local Government Competition Resulted In Serious Ecological Problems?

While promoting rapid development of China's economy, the local government competition focuses too much on economic interests and too little on environmental protection. Why did the local competition bring in so much damage to the environment in the past three decades? It's largely due to the imperfect system of local government competition. The local competition is based on political performance appraisal that takes no account of ecological costs. It is faced with little mechanism restraint, and there is little legally binding regulation in administrative decision-making procedure and accountability system. Under such a competition system, some local governments would rather pursue economic and social interests than govern the ecological environment.

Moreover, the lack of motivation for local governments to govern the ecological environment is also closely related to the spillover effect of ecological governance and the effect of "pollution haven". The spillover effect of ecological governance is determined by its nature, which is the process to provide ecological environment as public goods. Local governments may implement strict environmental standards, increase the penalties for illegal pollution, invest in building environmental protection facilities to govern the local environment. That is to provide the public with something that everybody can enjoy–sound ecological environment. Due to the non-exclusive and non-competitive nature of public goods, however, this is often accompanied by others taking a "free-ride". Ecological governance is no exception. The achievement of local ecological governance may spill over the administrative boundaries so that neighboring areas can enjoy the benefits of blue sky and clean rivers free of charge. This is the spillover effect of ecological governance. In order to win the local government competition, some local governments may wait for their "neighbors" to provide ecological "products" free of charge. The result might be that everybody's business is nobody's business.

While local governments are competing for capital, ecological governance would weaken local competitiveness, resulting in "pollution haven" effect. If local governments implement more stringent environmental standards, it is likely to force polluting enterprises to move to adjacent regions where environmental standards are relatively lower and because of such lower standards, those enterprises would unscrupulously pollute the environment. Then from a regional and national perspective, one single city with strict environmental policy may increase the total amount of pollution. Local governments need both motivation and capability to govern the environment well. And the current mismatch of administrative and fiscal powers on ecological governance between the central and local government would make things less efficient.

What Shall Be Done to Encourage Local Competition to Ensure Ecological Governance?

While competing for capital, local governments lack the motivation and ability to govern the environment. Due to the spillover and pollution haven effect, local governments would find it counterproductive in pooling resources for ecological governance. It may even increase the amount of total pollution. Does local competition inevitably lead to environmental damage? Not necessarily. Proper measures can be made to avert the spillover and pollution haven effect, encourage local governments to be more motivated and able to govern the environment.

First, make ecological governance by local governments more cost-effective and subsequent achievements more credible. The performance of ecological governance by local governments should be linked to improved investment environment, higher fiscal transfer payments and local government debt loans. Eco-environmental indicators should be incorporated into the investment environment evaluation system to guide investors to have a right attitude to the environment. Sound eco-environment should be viewed as an advantage in acquiring fiscal and capital support. The governance of environment should be as important as infrastructure building and government efficiency. In terms of fiscal transfer payment, the administrative and fiscal powers on ecological governance between the central and local government should be clearly defined. Ecological problems in one region should be solved by local governments. Cross-regional problems are coordinated by the central government and solved by relevant local governments.

Second, reform the performance appraisal system and the competition restraint mechanism, and encourage local governments to focus more on ecological governance than local competition. In the past, the GDP-based local government appraisal system mainly focused on such indicators as the growth rate of GDP, the amount of tax paid and foreign investment. This made local government officials indifferent to other things. Moreover, since decision-making process is not perfect, local government officials impose their will on regional economic and social development. More often than not, the economy is developed at the expense of environmental pollution. Therefore, the performance appraisal system should be reformed to take into account the opinions of the public and the environmental protection agencies. Such indicators as the number of letters and calls for ecological problems, the development status of environmental organizations, and environmental information disclosure should be added as well. More credible local government decision-making procedures and accountability systems should be established to encourage local officials to pay more attention to ecological governance.

Finally, a uniform environmental standard is to be implemented in relevant areas to avoid pollution haven effect. Regional joint control mechanism should be established to coordinate specific matters, adopt unified environmental standards and set up new trading system for regional pollutant emission. A monitoring system is to be built among local governments to curb and prevent such unpopular practice as lowering environmental standards for more investment. Local governments are expected to compete for the better not for the worse. Functional zoning system should be improved. Different regions are expected to develop their unique and strong industries. Therefore, environmental regulations and performance appraisal systems should be adjusted accordingly.

China has entered the post-industrial era. In accordance with the law of economic growth and international practice, green economy would be the driving force for economic growth against the backdrop of economic restructuring and industrial upgrading. As infrastructure improves and the mount of favorable policies decreases, the competition between local governments at all levels will gradually shift to technical innovation, improved ecological environment and so on. Proper measures could be made to guide local government competition to be more environmentally friendly and be dedicated to improving ecological governance.

第九章 以生态红线为基准谋划国土空间开发

随着经济发展和生态保护的双重诉求不断被提上发展议程，判别和保护国土空间的生态红线具有越来越重要的现实意义。生态红线保障了国土空间基本的生态安全，是可持续发展的生态保障。国土空间生态红线的实质是维护最低限度的生态安全格局，也就是维护生态系统服务最基本和最关键的土地和空间格局。

国土空间的开发利用必须以土地的生态过程和格局为依据，应优先进行生态红线即生态安全格局的控制，再根据经济社会发展需要进行建设用地的规划和布局。基于生态安全格局的城镇发展格局，即用尽可能少的土地，维护城市的基本生态系统服务，同时为城市发展提供充足的建设用地，是实现精明保护与精明增长，从而实现生态文明的有效途径。

确保生态安全　兼顾多种目标

生态基底为人类提供丰富完整的生态服务，满足人类的各项需求。因此生态安全格局针对城市发展的方方面面，从生态角度保证城市健康运行的基本安全，它主要涉及生态系统的气候调节、水源涵养、氧气供给等基本功能，以及退化生态系统恢复、生态系统自身健康维持、景观格局优化，以及对经济社会发展需求的满足等其他功能。多种功能往往不可能同时达到最大化，而且任何一种功能都或正或负地与其他功能相关，因此需要对主要功能进行权衡分析，实现多目标的协同发展，从而为相关生态规划的制定和生态系统的综合管理提供科学依据，最终更好地为人类谋福祉。

在确保生态安全的前提下兼顾多种目标，需要对生态基底条件有深刻的认识，同时对地区生态需求有精确的感知。在大数据分析技术、计算机处理技术、地理信息系统等日臻成熟的背景下，可以完成单项目标的最大化描述，形成水源涵养、洪水调蓄、荒漠化防治、水土保持、生物多样性保护等不同类型的安全格局，进而针对不同城市发展情景需求，模拟不同人口数量、发展目标等要求下生态效益的最大化方案，继而因应情景"整合"单一安全格局，形成综合的生态安全格局。作为国土空间、区域范围、城市空间生态安全的保障，生态安全格局提供了城市的基本生态系统服务，同时基础性地定义了城市发展方向，使得土地的生态过程和格局先行于城市的发展建设规划——优先进行不建设区域的控制，再根据经济社会发展需要进行建设用地规划和布局。这保障了发展建设不会"超纲越界"，增强了发展的基础安全性。

新常态下更应注重生态安全

1949年之后尤其是改革开放以来，我国城镇人口数量快速增加，生产力水平显著提高。总体来看，开发规模和强度的增大导致整体国土空间都呈现出生态承载能力下降、生态环境恶化的局面。同时，随着经济活动不断由东向西、由沿海向内地推进，发展战略在不断发生着转变，因此经济—生态环境的关系也随着发展方式的不同而不断变化：从改革开放后大力推进东南沿海地区经济发展，强调经济建设快速有效地进行，到"新常态"出现后着重强调"生态文明""绿色发展"，探索"经济"与"生态"协同发展模式，使其成为"带动内地发展""优化沿海发展"的新标杆。

在不同的发展阶段，国土空间开发对于生态环境有不同的需求，会由索取资源到恢复保育，再到耦合共赢。居民对生态系统服务有各异的要求，会逐渐由生存需求过渡到精神需求，再到和谐共存。新常态下迫切需要构建适合当前发展阶段的生态安全格局，从而既不妨碍经济发展需求，又能最大程度地降低对生态环境的破坏。通过构建评价指标体系，运用耦合度模型，基于区域经济发展综合指数与生态环境安全综合指数，可以对我国生态安全功能区经济发展与生态环境耦合度进行定量度量。随着城市化的不断加深，还应不断修正生态系统功能的新需求，不断改变经济—生态耦合的目标，从而不断调整生态政策以保证安全格局的与时俱进，保证国土空间利用的因时而动。

多策并举保障国土空间安全开发

在探讨生态红线即安全格局及国土空间集约高效利用方案的基础上，保障规划方案的顺利实施需要各社会阶层、各利益群体、各职能部门的协调配合。而协调配合的基础，应当以立法的方式将安全格局上升到"不可违背""不可逾越"的高度，在此基础上，需要专业职能部门的精确控制，同时需要全社会成员的共同参与和监督。

首先，将国土尺度生态安全格局纳入法定规划体系。在具体规划编制中，广泛征求各部门和利益相关者的意见，通过多方博弈最终确定其空间边界。生态安全格局的最终成果应该通过立法和相关政策实现永久性的保护，使之成为保障国土、区域和城市生态安全的永久性格局，并限制无序的城市扩张和人类活动。

其次，要有专门的管理部门负责区域生态安全格局的维护。为保证区域生态安全格局的有效实施，并统筹协调各个行业、不同地方政府在安全格局内的合理发展，应该成立专门的区域生态安全格局管理部门作为实施主体。管理部门可以通过一系列行政手段保障国土空间的安全开发，如制止为了短期的经济利益在安全格局内建设大型永久设施的行为，通过议事协调机构协调各个行业、不同地方政府的经济利益，并使其与生态利益相协调。

最后，结合全国主体功能区划推进国土空间安全格局的建立与集约开发。在禁止开发地区，推动自然保护区、湿地空间、森林公园、海洋空间等生态功能区的生态修复，提高对主要生态保护区的监管能力，保护区域内的生物多样性和珍稀动植物基因库。在控制限制开发地区，强化人口和产业相对集聚的城镇地区工业与生活环境的协调治理，加强对资源开采加工等产业活动的生态监管，积极推进资源开发地区的开发节奏、步骤、程度的控制方案实施，从而提高经济—生态协同发展的监控能力。

以生态安全为基础构建开发格局

国土空间作为经济社会发展的空间载体，其空间格局将直接决定开发建设的资源环境要素组合效率。生态安全格局是实现可持续发展的基础保障，有利于促进生态系统与社会经济系统的协调。在经济快速发展地区，构建生态安全格局的目标首先是保护生态系统的稳定性，并通过水平方向的有机链接，为经济的快速增长提供生态保障与环境支撑；在生态脆弱地区，其生态环境具有脆弱性和易变性的特点，加强区域景观建设和生态安全控制，有利于将景观演化导入良性循环；而在一些关键、重大工程的建设中，生态安全格局为重大工程的顺利进行和区域环境保护保驾护航。

国土空间开发格局的构建应以生态安全格局为基础。我国现行开发的区域特点是东南沿海长三角、珠三角、京津冀等城市群地区快速发展，资源占用比例高、生态环境压力大；内陆地区推进缓慢甚至长期停滞不前，经济发展相对缓慢，资源本底差，生态安全格局面临威胁，处于经济、生态的双重困境。因此须在"区域协调发展"方针的指引下培育内陆地区经济增长极，从而优化国土空间格局、转变国家经济布局与发展方式、积极应对世界经济格局调整。加强沿海—内陆的人流、物流、信息流的深刻联系互动，从而带动发展资源生产力要素由东向西、由南向北流动，同时平衡整体国土空间的自然资源占用。须"点、线、面"相结合，构建"城市群—发展轴—经济区"区域空间体系，合理配置各发展单元的功能，站在国土空间一体化的角度全盘考虑，增强区域发展协调性，利用竞争关系提升国土空间的竞争能力，利用协同关系促进国土空间的整体优化，形成国际化大都市—大城市—小城镇—农村，东南沿海、东北老工业基地、西南边陲、西北内陆地区、中部地区等各层级各单元各司其职、相互配合、共赢发展的国土空间开发格局。坚持先规划后建设的原则改革城市总体规划编制方法，以生态格局为基础合理划定城市开发边界。把握好城市定位，注重开发强度管控，实现城市开发边界和生态红线"两线合一",国民经济和社会发展规划、城乡规划、土地利用规划、生态环境保护规划等"多规合一"。杜绝对生态红线划定各自为政的现象,统筹各个部门根据业务范围划定的专项红线,实现一张图、一条线,统一目标和发展成果。

Chapter 9
Develop Geographical Space on the Basis of Ecological Red Line

As economic development and ecological protection are both in the spotlight, it is crucial to identify and protect the ecological red line of China's geographical space. Ecological red line guarantees basic land security and sustainable development. The red line is drawn to maintain the minimum level of ecological security.

The development and utilization of geographical space must be based on the ecological pattern of land. Priority should be given to control ecological red line, or the ecological security, before planning land use according to economic and social needs. It is an effective way to achieve smart growth and ecological governance by developing urban areas on the basis of ecological security, that is to say, using the least amount of land to maintain basic ecosystem services while providing sufficient land for urban construction.

Protecting Ecological Security and Achieving Multiple Goals

Eco-substrate provides rich and complete ecological services for human beings and meets their varied needs. Therefore, the ecological security pattern is aimed at guaranteeing basic security of urban functions from the ecological perspective. It covers a wide range of urban development, such as climate regulation, water conservation, oxygen supply, as well as the restoration of degraded ecosystems, the maintenance of the ecosystem per se, the optimization of landscape pattern and meeting economic and social needs. Many of those functions cannot be maximized simultaneously. One function may be co-related to another in a certain manner. It is therefore necessary to analyze major functions and achieve coordinative development. This can provide scientific evidences for ecological planning and comprehensive management of the ecosystem.

In order to achieve other goals under the premise of ensuring ecological security, it is needed to have a deep understanding on eco-substrate and precise perception of regional ecological needs. Many advanced technologies, such as big data analysis, computer processing and geographic information system, are developing by leaps and bounds. It is therefore possible to maximize the effect of one goal, and form different types of security pattern, including water conservation, flood regulation, desertification control, soil and water conservation and biodiversity conservation. Furthermore, it is possible to simulate the number of population and development goals in different cities and draw a blue print to maximize the ecological benefits. A single security pattern could then be integrated, including a comprehensive ecological security pattern. The ecological security pattern, which guarantees the security of geographical space, including regional and urban space, provides basic urban ecosystem services, charters the course for urban development, and ensures that the ecological pattern of land precedes urban development planning. It means that priority is given to control land not for construction before planning land for construction according to economic and social needs. This can make sure that construction will not cross the bottom line and development could be more sustainable.

Paying More Attention to Ecological Security Under the New Normal

Since in 1949, in particular the reform and opening up, China's urban population is increasing rapidly and its productivity develops by leaps and bounds. The increasing scale and intensity of development leads to the deterioration of ecological environment. As the economy is developing from the east to the west and from the coastal cities to the hinterland, the development strategy, as well as the economy-environment relationship, should also be changed. Since the reform and opening up, the focus has been on the rapid and effective economic development in eastern coastal regions, but when the economy enters the New Normal era, the focus should be on green economy, which explores a coordinated development between the economy and ecology. This new focus aims to optimize economic development in coastal cities while upgrade economic growth in the hinterland.

In different stages of development, the development of geographical space requires different needs for ecological environment. It evolves from the exploration of resources, restoration and conservation, to a win-win scenario. Residents would have different requirements for the ecosystem services, from the survival needs, to the spiritual needs and finally coexistence. Under the New Normal, it is urgent to construct an ecological security pattern suitable for the current development stage, which will boost economic growth and minimize environmental damage. It is possible to have quantity analysis on the coupling degrees of China's economic development and ecological environment in eco-security regions through the establishment of the evaluation index system using the coupling model and comprehensive indicators for regional economic development and ecological security. As urbanization expands, it is advisable to constantly revise the new needs of ecosystem and change the economic-ecological coupling goals. It is necessary to adjust environment policies to keep security pattern up-to-date and use the geographical space accordingly.

Various Measures should be Adopted to Ensure Safe Exploration of Geographical Space

The blue print is drawn after discussion on how the security pattern and geographical space can be intensively and efficiently utilized. Smooth implementation of the blue print requires coordinated efforts by people from all walks of life and different authorities. The security pattern should be made into law that is unbreakable. The law would be the foundation for coordination.

First, the ecological security pattern should be incorporated into the legal system. During the legislation process, opinions from a wide range of authorities and stakeholders should be consulted to determine its spatial boundaries. The final outcome of the ecological security pattern should be permanently protected by laws and related policies. The outcome should be a permanent pattern that safeguards geographical space, regional and urban ecological security. It can also inhibit disorderly urban expansion and human activities.

Second, there must be a dedicated management authority responsible for the maintenance of regional ecological security pattern. A special authority should be established to ensure the effective implementation of the regional pattern and to coordinate rational development of various industries with different local governments in the region. The authority can ensure safe development of geographical space through a series of administrative measures, such as stopping the construction of large permanent facilities within the region for short-term economic interests. It is able to coordinate the economic interests of various industries with local governments through the coordinating bodies in line with the ecological interests.

Finally, the security pattern should be developed in light of major functional zoning areas. It is advisable to encourage ecological restoration of such functional areas as natural reserves, wetland spaces, forest parks and marine spaces where development is prohibited. More efforts should be made to regulate major ecological reserves and protect biodiversity and rare animal and plant genes in the region. In restricted areas, more efforts should be made to coordinate industrial development with living environment, and monitor such industrial practices as exploration and processing of resources. The rhythm, steps and procedure to exploit resources should be controlled in order to better monitor economic-ecological development.

Reshaping Development Pattern Based on Ecological Security

It is in geographical space that economic and social development can be advanced. The space pattern determines how efficient resources are used. The ecological security pattern is the basic guarantee for sustainable development, which is conducive to the coordination between the ecosystem and the socio-economic system. In the fast-growing areas, the goal of forming the pattern is to protect the stability of the ecosystem and provide ecological and environmental support for the rapid economic growth through the organic links in the horizontal direction. In the ecologically fragile areas whose ecosystem is fragile and volatile, it is advisable to strengthen regional landscape construction and ecological security control, which leads landscape evolution to a virtuous circle. For the construction of major projects, ecological security pattern guarantees smooth progress and protects regional environment.

The construction of geographical space pattern should be based on ecological security pattern. The characteristics of China's current development can be summarized into two aspects. First, the city clusters in the Yangtze River Delta, the Pearl River Delta, the Beijing-Tianjin-Hebei regions are developing rapidly, which accounts for high proportion of resources and puts heavy pressure on the ecological environment. Second, inland cities are growing slower and resources are not efficiently used, which threatens the ecological security pattern. The economy and ecology are both in troubles. It is therefore necessary to cultivate new engines for economic growth in inland cities under the guidance of "regional coordinated development", so as to optimize geographical space pattern, optimize the overall economic layout and development mode, and respond to the adjustment of the world's economic structure. It is necessary to strengthen the connectivity of people, goods and information between coastal and inland cities, lead the flow of resources from the east to the west and from the south to the north, and balance the use of resources. The dots should be connected to build a system comprising city clusters, development axis and economic zones. Different regions should give full play to their advantages and improve the competitiveness of geographical space. It is advisable to build international metropolis, big cities, small towns and villages in southeast coastal regions, northeast old industrial bases, southwest borders, northwest inland areas and the central provinces in accordance with local conditions. City planning should precede urban construction. The urban space line should be drawn in light of ecological pattern. Cities should pay attention to control the development pace in accordance with the ecological red line. Balance should be reached in the planning of economic growth and social progress, urban and rural development, land use and environmental protection. Different authorities should plan and act according to the same ecological red line toward the same goal.

第十章　生态文明建设要算好自然资源这本账

十八届三中全会提出"探索编制自然资源资产负债表，对领导干部实行自然资源资产离任审计，建立生态环境损害责任终身追究制"。编制自然资源资产负债表实际上是核算自然资源资产的价值及其变化，摸清"生态家底"，全面反映经济发展的资源环境代价和生态效应。自然资源资产负债表是自然资源资产的"记录仪"，是资源精细管理与生态风险管控的"对账本"，是判定生态文明建设成效的"裁判员"，是倒逼经济发展模式转变的"推进器"，对于生态文明建设，将起到基础性数据支撑、关键性决策依据、科学性成果判定的作用。

建立资源资产账本，摸清"生态家底"

自然资源是指一切天然存在且有利用价值，可提高人类当前和未来福祉的自然环境因素。自然资源资产负债表是指用资产负债表的方法，将全国或一个地区的所有自然资源分类并进行价值核算而形成的报表。它通过核算自然资源资产的存量及其变动情况，全面记录当期（期末—期初）自然和各经济主体对自然资源的占有、使用、消耗、恢复和增值活动，进而评估当期自然资源实物量和价值量的变化。因此，自然资源资产负债表的直接功能是为了建立自然资源资产的"账本"，摸清"生态家底"及其变动情况。

对自然资源进行核算的历史可以追溯到20世纪70年代。1978年，挪威最早开始了资源环境核算，其环境账户以国民经济投入产出为模型，为决策者评估能源交替增长提供参考依据。随后芬兰、法国、美国等国家先后进行了自然资源核算的实践。20世纪90年代以来，依据可持续发展的理念，联合国、欧盟委员会、国际货币基金组织、经济合作与发展组织、世界银行等组织致力于环境与经济综合核算体系（System of Environmental and Economic

Accounting，SEEA）的理论研究和实践应用，相继发布了 SEEA（1993）、SEEA（2003）、SEEA 中心框架（2012）等重要报告，制定出环境—经济账户的国际统计标准，正式把自然资源与环境等要素纳入国民经济账户。2012 年后，各国都在 SEEA 中心框架下结合本国实际探索编制具有本国特色的自然资源资产负债表。

我国的自然资源核算研究始于 20 世纪 80 年代中后期，有学者翻译国外关于自然资源核算的研究报告，开展相关课题研究，并发表系列文章，呼吁开展资源核算工作。1992 年联合国环境与发展大会通过的《21 世纪议程》要求"在所有国家中建立环境与经济一体化的核算系统"。随后，"建立一个综合的资源环境与经济核算体系"被列入《中国 21 世纪议程》。2001 年，国家统计局以重庆市为试点，开展资源环境核算，并于 2003 年试编《全国自然资源实物量表》。2004 年，国家林业局和国家统计局联合开展中国森林资源核算。国家统计局与国家环保总局于 2006 年联合发布《中国绿色国民经济核算研究报告 2004》，其指出 2004 年因环境污染造成的损失占当年 GDP 的 3.05%。

近年来，一些地方开始了探索编制自然资源资产负债表的实践。2014 年 5 月，贵州省率先将编制自然资源资产负债表列入地方性法规。2015 年，深圳市以生态保护为重点的大鹏新区和以工业发展为重点的宝安区，推出了我国首个县区级自然资源资产负债表；浙江湖州市以森林资源和水资源为重点，河北承德市以森林资源为重点，陆续推出了市级自然资源资产负债表。2015 年，国家在内蒙古呼伦贝尔市、浙江湖州市、湖南娄底市、贵州赤水市、陕西延安市等 5 地开展了编制自然资源资产负债表试点工作。

助力生态治理与发展转型

作为以制度保护生态环境、建设生态文明的战略举措，自然资源资产负债表能够摸清自然资源资产的"家底"，全面反映某一时期自然资源资产负债的变化情况，为自然资源的高效配置提供数据支撑；明确自然资源资产所有者权益和管理者的责任,为领导干部自然资源资产离任审计及政府生态环境绩效评价提供量化依据;促进改善环境治理和生态修复制度,倒逼经济发展模式向以绿色 GDP 衡量的可持续发展方式转变。

第一,厘清自然资源资产的存量、变化及负债情况,为自然资源有效配置和利用提供基础。通过记录自然资源资产的数量与质量以及生态资产的占用、消耗、恢复和增值情况,自然资源资产负债表可以为自然资源资产管理提供数据支撑,揭示各类主体对自然资源的占有使用情况及负担程度,评估区域环境承载力和生态环境风险,有助于建立和实施自然资源有偿使用制度、生态补偿制度和环境监测预警制度等,使自然资源得到更加高效的配置和利用。

第二,为地方政府及其领导干部的政绩考核提供参考依据,引导地方政府治理生态环境。在编制自然资源资产负债表的基础之上,可以将对地方政府的考核与本地区自然资源资产负债差额挂钩,并作为评价地方政府治理成效的重要标准,推进资源环境由被动管理转向主动治理。资产负债表还可以成为领导干部实行自然资源资产离任审计的依据,若官员任期内监管不力而造成自然资源资产贬值过多,将不被提拔甚至被问责,促进各级领导干部树立正确的政绩观。

第三,扭转资源消耗大、环境污染严重的传统发展方式,推动经济结构与发展模式的转型。编制自然资源资产负债表后,投入产出分析就不仅限于直接投入的土地、矿产、水等资源,而是推广到一切与之相关的自然资源及其变动情况。资源损耗、环境污染、生态损害都将削减生态资本,从而缩减社会财富。因此,出于增加净资产减少负债的考虑,政府将会裁减自然资源消耗高的工农业而增加高新技术产业,以增加绿色 GDP,从而实现经济结构的转型和发展方式的转变。

攻克编制难点,算好自然资源资产账

从各地编制自然资源资产负债表的实践来看,编制工作存在着诸多难点。

首先,支撑资产负债表的基础统计数据难以全面、准确、及时获取。自然资源包括水、土地、森林、矿产和野生动植物等众多领域,而且生态环境系统复杂,对这些自然资源资产进行评估统计比较困难。已有的自然资源资产统计制度不够系统、完善,统计工作手段滞后,基础数据既无法保证质量,又难以确保时效。一些地方的编制实践表明,除土地资源数据外,其他数据长期缺失、时效性差、难以快速收集。

其次,自然资源资产价值化的规范模式尚未统一。定价有利于不同资源、不同地区之间的统一比较。但鉴于自然资源资产种类众多、用途广泛、因素繁杂,价值随着生产力的变化会发生巨大改变,还须考虑资源消耗、环境退化、稀缺性等诸多因素,自然资源资产的价值评估工作非常复杂。成本法、市场法、收益法、意愿法各有其优缺点,不管是理论界还是实务界在选择自然资源资产估价方法时很难达成共识。

再次,负债内涵外延不明确。有观点认为,负债是自然资源被开采、耗费后减少的价值。另有观点主张负债应该被定义为把自然资源维持在某个规定水平之上的成本。地方试点对于负债内容的实践也不尽相同。浙江湖州市将负债项定为资源耗减及其造成的环境损害,而河北承德市则将资源过耗、环境损害和生态破坏三个部分列入负债核算。

最后,编制过程的协同共享机制尚未健全。虽然国家试点文件确定统计部门为牵头单位,但是统计部门与水利、林业、矿产、土地、环保、规划等自然资源管理部门之间仍存在指标口径、分类标准、数据共享等方面的不统一。许多地方编制资产负债表的过程中,往往是将资产负债表交由国土局、环保局、水利局、水文局等相关部门填写,缺乏统筹协调机制和监管办法,而且还存在一些没有部门过问的"空白地带",严重影响资产负债表数据的一致性、准确性和全面性。

为了克服编制中存在的困难,科学规范地编制自然资源资产负债表,可以从如下几个方面着手。

第一,加大专项投入力度,创新数据可得技术,完善自然资源资产统计制度。明确自然资源资产的统计对象、范围、频率、方式,确保统计的基础数据质量过关，加强大数据、卫星遥感、地质勘探、地理信息系统建模等创新技术方法的应用，增加自然资源资产的实物数据的可得性，同时进行动态监测，把握特定时间、空间自然资源的变化。

第二，价值化采取循序渐进、因"类"制宜的策略予以推进。对特定区域、容易定价的自然资源，在充分考虑自然资源的经济、生态和社会价值的基础上，可以根据影子价格模型、CGE模型、边际机会成本模型、市场估价模型等定价方法予以确价。对于不同种类的自然资源采取不同的经济模型予以价值化，根据自然资源市场交易的发展状况，借鉴已成功定价资源的经验，逐步推广。以水资源为例，不同类的水质量不同，使用价值和生态价值亦存在差异，相应地，I、II、III类水和IV、V类水可以分别根据饮用水水源地取水价格和工农业用水资源费来定价。再如，矿产资源可以根据其消耗成本定价。

第三，统一负债概念，明确负债类别。一致的负债概念有利于规范自然资源资产负债表的编制。确定负债含义时，不仅须考虑到现有已发生的自然资源资产价值减损或功能下降，同时也应将未来有可能继续发生的损耗和退化、治理所需要的投入成本纳入其中。至于负债的类别，则应该在进行基础分类的同时，针对自然资源资产不同的损耗情况，进行多级分类，以全面反映地区自然资源利用及其社会影响状况。

第四，健全不同部门之间的协同共享机制。国家统计局应当尽快出台关于自然资源资产的统一指标、分类及数据标准等文件，避免部门间继续出现数据、指标等难以匹配的问题，搭建跨部门的信息协调及工作协同的平台，协调各相关部门的职能，为全面开展自然资源资产核算和自然资源资产负债表编制提供有力的组织保障。

Chapter 10
Building an Ecological Civilization Entails a Clear Understanding of Natural Resources

The Third Plenary Session of the 18th CPC Central Committee proposed measures of "exploring the compilation of nature resource assets balance sheet, auditing leading cadres regarding natural resource assets when they leave offices and establishing a lifelong accountability system for eco-environment damages". Compiling natural resource assets balance sheet can actually account for the value of natural resource assets and their changes, get a clear picture of ecological resources and comprehensively reflect the recourse and environmental cost and ecological effect of economic development. These types of balance sheets are "recorders" of natural resource assets. "reconciliation books" for lean management of resources and ecological risk control, "judges" for eco-civilization construction results and "propellers" to force the transformation of economic development mode. They can provide basic data as key evidence for making decisions and evaluating ecological achievements in a scientific way.

Establishing Account Books for Resource Assets to Have a Clear Picture of All Ecological Resources

Natural resources refer to all natural and environmental factors of value that can be used to improve the current and future well-being of mankind. Natural resource asset balance sheets mean that all natural resources in a region or in a country are classified into accounting statements after value calculation in the form of balance sheets. They can full record the occupation, use, consumption, recovery and value-added activities of natural resources by all stakeholders in a certain period (from beginning to end of the period) and then assess the changes of physical quantity and quantity of value of natural resources in that period through calculating the stock and changes of natural resource assets. Therefore, the direct function of the natural resource assets balance sheet is to establish account books for resource assets and understand the quantity of resources and their changes.

The history of natural resource accounting can be dated back to 1970s. In 1978, Norway was the earliest country to begin environmental accounting. Using national economic input as models, Norwegian environmental accounts provided references for decision-makers to evaluate energy increase. Later on, Finland, France, USA and other countries began to follow suit. Since 1990s, in line of the concept of sustainable development, UN, European Commission, IMF, OECD, WB and other organizations were committed to the theoretical research and practical application of System of Environmental and Economic Accounting (SEEA) and issued SEEA (1993), SEEA (2003), SEEA Central Framework (2012) and other important reports, developing an unified international standard on environmental and economic accounts and formally integrating natural resource and environmental factors into national economic accounts. After 2012, all the countries began to compile natural resource asset balance sheets with their own characteristics in accordance with their specific national conditions under SEEA Central Framework.

China's natural resource accounting research started in middle and late 1980s when scholars translated foreign research findings on natural resource accounting, conducted relevant researches, published a series of articles and called for the work of resource accounting. *Agenda 21* passed by United National Environment and Development Summit in 1992 required that an integrated environmental and economic accounting system be set up in all the countries. Later on, this requirement was included into *Agenda 21 for China*. In 2001, National Bureau of Statistics (NBS) began to conduct environmental accounting in Chongqing Municipality as a pilot project and compiled *National Natural Resource Physical Scale* in 2003. In 2004, State Forestry Administration and NBS jointly conducted forestry resource accounting in China. In 2006, NBS and Ministry of Environmental Protection jointly issued *Report on China Green National Economic Accounting Research 2004*, which pointed out that economic losses caused by environmental pollution in 2004 accounted for 3.05% of the year.

In recent years, some places have begun to compile natural resource assets balance sheets. In May, 2014, Guizhou Province took the lead in including the compiling of natural resource assets balance sheets into local regulations. In 2015, Shenzhen launched China's first county-level natural resource asset balance sheets in Dapeng New District focusing on ecological conservation and Bao'an District focusing on industrial development. Huzhou City, Zhejiang Province and Chengde City, Hebei Province launched their balance sheets focusing on forestr and water resources (Huzhou) and forest resources (Chengde) respectively at city level. In 2015, China implemented similar pilot projects in Hulunbeier City of Inner Mongolia, Huzhou City of Zhenjiang Province, Loudi City of Hunan Province, Chishui City of Guizhou Province, Yan'an City of Shaanxi Province.

Boosting Ecological Governance and Developmental Transformation

As a strategic measure to protect the ecological environment and build ecological civilization, compiling natural resource assets balance sheet can help to know the quantity of natural assets, fully represent the changes of natural assets in a certain period and provide data support for efficient allocation of natural resources. It can also clarifies the rights and interests of the owners of natural assets and the responsibilities of managers, which can provide quantified evidence to audit natural assets when officials in charge leave their posts and to evaluate ecological and environmental perforce of governments. The measure can also promote the improvement of environmental governance and ecological restoration systems and change the mode of economic development to sustainable development measured by green GDP.

Firstly, understanding the stock, change and liabilities of natural assets can provide good bases for efficiently allocating and utilizing natural resource. Through recording the quantity and quality of natural assets and the occupation, consumption, restoration and value appreciation of natural assets, the balance sheets can provide managers of natural assets with data support and reveal the stakeholders' ownership of natural assets and their ecological burdens so as to evaluate the carrying capacity of regional environment and ecological risks. They are also conducive to establishing and implementing the systems of paying for the use of natural resources, of ecological compensation and of environmental surveillance and early warning, etc., so that natural resources could be allocated and utilized more efficiently.

Secondly, compiling the balance sheets can provide reference for the evaluation of performance of local governments and local leaders and motivate local governments to conduct effective eco-environment governance. On the basis of compiling the balance sheets, we can directly link performance evaluation of local governments with assets and liabilities of natural resources and use the balance sheets as an important standard to assess the effects of local ecological governance, which can promote the change of governance modes from passive management to active governance. Balance sheets can also be the basis for auditing leading officials regarding natural resource assets when they are transferred to other posts. If the officials have conducted inadequate supervision during their terms, which has caused too much depreciation of the natural assets, they will not be promoted or even held accountable. It will motivate leaders at all levels to develop a correct concept on political performance.

Thirdly, compiling the balance sheets can help to reverse the traditional ways of development which consume tremendous resources and cause serious environmental pollution and promote the transformation of economic structure and development modes. With these balance sheets, input-output analysis will not be limited to direct input of land, minerals, water and other resources. Rather it will cover all the natural resources related to economic activities and their changes. Resource loss, environmental pollution and ecological damages will all reduce ecological capital, thereby reducing social wealth. Therefore, for the sake of increasing net assets and reducing liabilities, the governments at all levels will downsize industry and agriculture with a high consumption of natural resources and develop more high-tech industries to increase GDP so as to achieve the transformation of economic structures and development modes.

Overcoming the Difficulties in the Compilation and Keeping a Good Account of Natural Assets

There exist many difficulties in the compilation work in some places.

Firstly, it is difficult to obtain comprehensive, accurate and timely basic statistical data supporting the compilation. Natural resources include water, land, forest, minerals, wild animals and plants and many other categories. Furthermore, great difficulties exist in evaluating these natural assets within an complex eco-environmental system. The current natural assets statistical system is not systematic and complete and statistical methods are lagging behind, so the quality and timeliness of basic data cannot be guaranteed. Compilation practices in some places indicate that data except land resource data are missing or are not timely and difficult to collect rapidly.

Secondly, there has been no unified standard in defining the value of natural assets. A standard pricing method is conducive to comparing different resources in different regions. However, values of assets will undergo tremendous changes with the changes of productivity due to different types, uses and factors of resource assets. Also many other factors such as resource consumption, environmental degradation and scarcity need to be taken into consideration. Assessing the value of natural resource assets are very complex and methods like cost method, market method, benefit method, intention method, etc. have their own advantages and disadvantages. Researchers and officials find it very hard to reach consensus on the selection of natural resource assets valuation methods.

Thirdly, the connotation and denotation of liabilities are not clear. One point of view holds that liabilities are the subtracted value after natural resources are exploited and consumed. Another point of view believes that liabilities should be defined as the cost that keeps natural resources above a certain specified level. The contend of liabilities is defined differently in different pilot projects. Huzhou City of Zhejiang Province defines liabilities as resource reduction and environmental damagers while Chengde City of Hebei Province lists resource over consumption, environmental damage and ecological disruption into liabilities accounting.

Last but not least, the collaborative and sharing mechanism in the compilation process has not been established. Although the official pilot guidelines identified statistics departments are the leading agencies, there have been discrepancies on indicator scopes, classification standards and data sharing, etc. between statistics departments and departments managing natural resources such as departments responsible for water conservancy, forestry, minerals, environmental protection and planning, etc. In the compilation process, many regions simply ask departments of land resources, environmental protection, water conservancy, hydrology, etc. to fill in the balance sheets due to lack of a unified system of coordination and supervision. Also there are some "blank zones" which no department is concerned with. This has seriously affected the consistency, accuracy and comprehensiveness of data in the balance sheets.

In order to overcome difficulties in the compilation process and compile the balance sheets in a scientific and standard way, the following measures can be taken:

Firstly, we should enhance special input, innovate data availability technology and improve statistical system on natural resource assets. We should clearly define the statistical targets, scopes, frequencies and modes of natural assets to ensure the quality of basic data, strengthen the application of innovative technologies such as big data, satellite remoting sensing, geological prospecting and geographic information system modelling, etc. and enhance the availability of physical data. Meanwhile, we should conduct dynamic surveillance and capture the changes of natural resources at a specific time and place.

Secondly, we should adopt the step-by-step and category-based strategy to promote valuation of natural assets. For natural resources in a specific region which can be easily priced, we should determine the price using such methods as shadow price models, CGE model, marginal opportunity cost model, market evaluation modelling, etc. on the basis of full consideration of the economic, ecological and social values of natural resources. Different economic models will be used to price different categories of natural resources. The valuation and pricing methods will be promoted gradually in line with the development of natural resource market transactions and successful experiences of others. Take water resources as an example. The quality of different types of water varies and the utilization value and ecological value are different, too. Accordingly, water prices of Category I to V can be determined based on the prices at water sources and also resource fees of industrial and agricultural waters. Another example is mineral resource, which can be priced according to the cost of consumption.

Thirdly, we should unify the concept of liabilities and clarify liability types. A consistent concept of liability is conducive to standardizing the compilation of the balance sheets. When we define liabilities, we should consider not only the depreciation of natural assets and the decreases of functions, but also the possible depreciation and degradation of resources in the future and the treatment cost. We should also classify different types of liabilities into different layers based on the consumption of natural assets in order to fully reflect the regional use of natural resources and its social impact.

Fourthly, we should enhance the coordination and sharing mechanisms between different departments. NBS should promulgate unified indicators, classifications and data standards, etc. regarding natural resource assets to prevent the discrepancies in data and indicators prepared by different departments. A cross-department platform for information coordination and work sharing mechanisms will be launched so as to coordinate functions of different departments, which can also provide strong and institutional guarantee for the officials so as to carry out comprehensively the natural resource assets accounting and compilation.

第十一章　践行生态补偿　谋划绿富双赢

人类的经济、社会系统都是建立在生态系统上，并与其"共存共荣"。生态系统良性运转是人类社会可持续发展的前提。长期以来，生态环境问题之所以难以解决，在很大程度上是因为受益者、保护者、破坏者、受损者之间的经济、生态利益未能实现公平分配。生态补偿正是根据生态系统服务价值、生态保护成本、发展机会成本等因素，综合运用行政、经济、法律手段，形成生态破坏者赔偿、受益者付费、保护者与受损者得到合理补偿的运行机制，以实现生态利益的分配正义和生态责任的公平承担。完善生态补偿制度，对于实施主体功能区战略，促进生态扶贫，谋划绿富双赢，使"绿水青山"真正成为"金山银山"具有重要意义。

生态补偿的实践与探索

在国际上，生态补偿相对应的称谓是"生态服务付费"（payment for environmental/ecological services）。总体而言，生态补偿的国际实践主要分为两种模式：第一是政府主导模式，包括生态补偿基金、生态补偿税、区域转移支付和流域合作等。该模式的主要特征为政府购买与主导、由政府机构管理、实施范围大、目的多样化、支付方式和标准统一及效率较低。例如美国的耕地保护性储备计划（CRP）、墨西哥水文环境服务支付项目（PSAH）和英国环境敏感地带项目（ESA）、农村管理计划（CSS）等。第二是市场主导模式，包括绿色偿付、配额交易、生态标签体系、排放许可证交易和国际碳汇交易等。该模式的主要特征为生态服务使用者或受益者付费、自愿性、由中介机构管理、实施范围小、目的单一、补偿方式多样、效率较高。例如厄瓜多尔的 PROFAFOR 碳封存项目、玻利维亚的 Los Negros 流域保护项目和法国 Vittel 水保护项目。这些生态补偿的实践由于法律规范健全、政府支付能力较强、产权制度完善、市场机制成熟、多方主体参与等因素都取得了一定成效。

自20世纪七八十年代起，我国四川、云南等省就开始了生态补偿的实践。从20世纪90年代中后期开始，在森林、草原、矿山等多个领域开展试点，生态补偿实践得以快速推进。四川、陕西、甘肃三省率先开展退耕还林工程试点，2002年扩大到20多个省区。广东、福建等省份开展了流域生态补偿试点。近年来，各地试点在广度和深度上有了很大提升。深圳市于2007年开始对大鹏半岛开展生态补偿，通过财政转移支付的方式，对大鹏半岛近1.7万的原村民发放生态保护专项基本补助，补助数额达每人每月1000元，激发了原村民的积极性，使大鹏新区的生态环境得到改善。2011年，在财政部、环保部的牵头下，浙江和安徽在新安江流域开展了为期三年的全国首个跨省流域生态补偿试点。中央及此两省共同出资，根据年度水质是否达到考核标准来确定资金的拨付。新安江流域三年总体水质为优，符合水质考核要求。近年来，九洲江、汀江—韩江、东江等跨省流域横向生态补偿也逐步推进。这些实践有助于实现区域发展过程共赢、发展机会公平、发展成果共享。

生态补偿推进的制约因素

虽然我国在森林、草原、湿地、流域等领域及重点生态功能区的生态补偿方面积累了丰富经验，取得了初步成效，但专项法律缺失、补偿力度不足、配套制度不健全等问题逐渐凸显，严重制约了生态补偿的推进。

第一，生态补偿专项法律缺失，现有规范过于原则、分散。一是缺乏生态补偿的专门立法。虽然《生态补偿条例》在2010年就被列入立法计划，但因所涉及的利益关系复杂，补偿原则、领域、主体、资金、方式、对象、绩效评估等核心内容尚在探索之中，导致条例迟迟未能出台，使得生态补偿难以得到有效、全面落实。二是现有生态保护规范过于原则、分散，不成体系，难以形成制度合力。《环境保护法》《草原法》《森林法》《矿产资源法》等法律法规仅有关于生态补偿的原则性规定，可操作性差；地方法规、规章多集中在森林、矿产领域，覆盖面窄，且效力位阶低下，难以为生态补偿提供充分的法律依据。

第二，生态补偿范围窄、标准低、方式少、资金渠道单一且监管不足。一是现有生态补偿主要集中在森林、草原、矿产资源开发等领域，流域、湿地、海洋、土地、荒漠等生态补偿尚处于起步阶段，大气等领域尚未纳入。二是目前的补偿标准普遍偏低，不能完全弥补生态保护成本和发展机会成本，遑论生态系统服务价值。三是生态补偿方式单一。目前，生态补偿主要以资金补助为主，产业扶持、技术援助、人才支持、就业培训等政策、实物、技术及智力补偿方式未得到重视和推广。四是生态补偿资金渠道单一且监管不足。目前资金主要来源于中央财政转移支付，区域之间、流域上下游之间、不同社会群体之间的横向生态补偿发展不足，生态产品的市场交易机制很不健全。部分地区对生态补偿资金的使用并未到位，挤占、挪用资金的现象仍有发生。

第三，产权制度、监测体系、计算方法等基础支撑内容亟待完善。一是产权制度不健全。产权清晰是分配受益、认定补偿、确认权责的前提，而我国自然资源资产产权制度尚存在主体缺位、边界模糊等问题。产权制度不完善将会导致资源的掠夺性使用，受益者与保护者、破坏者与受损者的权责边界模糊，难以实现生态效益的公平和有效分配。二是监测体系不完善。当前，生态监测职能分散，监测网络割裂，监测方法多样，监测指标各异，监测人员、技术、经费等能力建设不足。三是计算方法不统一。机会成本法、意愿调查法、市场法等确立生态标准的方法尚未形成共识，影子工程法、当量法、市场价值法等多种计算生态服务价值的方法也存在争议。

推进生态补偿势在必行

生态补偿实质是实现发展成果的再次分配和资源蛋糕的重新切割，是促进经济转型、社会公平、生态优化协调统一的有效方式，必须坚定不移地推进生态补偿。

第一,立法先行是生态补偿规范化的根本保证。一要尽快出台《生态补偿条例》及其实施细则和技术指南。顶层设计上要明确生态补偿的原则、领域、对象、标准、程序、资金来源、考核机制、责任追究、主体权利义务等内容,为生态补偿的规范运行搭建完整的法律框架,提供完善的制度保障。二要修订相关法律法规,使之与生态补偿专项立法衔接得当,形成水、土、林、草、海、荒漠、湿地等全方位、多层次、广覆盖的生态补偿机制。

第二,补偿标准、范围、资金来源、方式的多元化是生态补偿全面实施的关键所在。一要建立系统、动态的生态补偿标准。针对草原、森林、矿产、荒漠、流域等领域及重点生态功能区的各自特点,分别制定生态补偿标准,形成系统的生态补偿标准体系。实施期间,根据保护效果、政府财力、社会经济等情况变化,适时调整生态补偿标准。二要扩大适用范围、拓宽资金渠道、丰富补偿方式,实现多元化的生态补偿机制。在森林、矿产、草原、流域等领域取得的试点经验基础上,逐渐推广应用到大气、湿地、重点生态功能区,以实现生态系统的整体保护。加大财政支持力度,优化财政支出结构,完善排污权、水权、碳汇交易等市场化补偿机制,鼓励及引导民间资本的参与,逐步建立政府引导、市场推进、社会参与的多元投融资渠道。除了资金补偿,对口协作、产业转移、人才培养、共建园区等受益地区向保护地区所提供的多样化横向补偿方式应当大力提倡。多元化补偿方式有利于培养被补偿地区的"换血"和"造血"功能,有助于贫困地区得到财政补偿、发展生态经济、建设基础设施、培养相关人才、保护生态环境从而脱贫致富。

第三,健全产权制度、完善监测体系、强化技术支撑是生态补偿有序、精确、高效开展的有力保障。一要健全自然资源资产产权制度。生态补偿所采用的不只是行政规制手段,还有市场运行方式,对公共资源最优配置的前提就是明晰产权。主体权利义务均衡、产权边界清晰、权能完整、利益分配公平的产权制度有利于开展生态补偿。二要完善生态监测体系。通过健全生态监测机制,规范监测方法,完善监测指标,利用智能化、规范化的监测技术形成统一的监测网络,加强监测合力,提供完整、一致、动态的监测信息。三要强化技术支撑体系。增加地理信息技术等研究及应用力度。如通过自动遥感分类工具,明确资源利用格局的变化,形成生态资产清单,构建自然资源资产负债表,实现单位面积价值量与生态补偿挂钩的精细化管理。明确并统一生态标准和计算生态服务价值核算方法,积极探索定量化的生态补偿评价体系。

第四,强化过程监督、健全绩效评估、调动社会参与是生态补偿切实推进的坚实后盾。一要强化监督追责。建立生态补偿协调机制,量化各部门的权责,加强对重点领域和区域资金分配使用情况的监督检查力度,对于落实不力的情况,启动监督问责机制。二要健全绩效考核评估。明确定期对生态补偿进行动态考核,对于践行良好的地区,增加政策倾斜和资金奖励;对于落实不力的地区,减少政策扶助和资金支持。三要调动社会参与。加大生态补偿宣传教育力度,构建公开公正的补偿程序,充分发挥媒体的监督作用,提升公众支持生态补偿的意识,引导各类受益主体履行生态补偿义务,督促管理者切实履行生态治理责任。

Chapter 11
Win-win: Ecological Compensation for Wealth and Green Development

Mankind's economic and social system is built on the ecosystem, and there should be peaceful coexistence and common prosperity between the two systems. The healthy operation of the ecosystem is the prerequisite for the sustainable development of human society. For a long time, ecological and environmental problems have been difficult to solve, which, to a large extent, is due to the fact that the ecological and economic interests of beneficiaries, protectors, polluters and victims are not fairly balanced. Based on service value, ecological protection cost and opportunity cost of ecological system and other factors, ecological compensation comprehensively uses administrative, economic and legal means to establish an operational mechanism in which polluters make compensations, beneficiaries make payment and protectors and victims get reasonable compensation so as to distribute ecological interests fairly and shoulder ecological responsibility equitably. Improving ecological compensation mechanism is of great significance in implementing the strategy of main functional areas, promoting ecological poverty alleviation, achieving win-win in both wealth accumulation and green development so that green waters and blue mountains can truly turn into mountains of gold and silver.

Practice and Theory on Ecological Compensation

Internationally, the corresponding name of ecological compensation is payment for environmental/ecological services. In general, there are two models of ecological compensation in international practice: the first one is government-led model which includes ecological compensation funds, taxes, regional transfer payment and river basin cooperation, etc. The main features of this model are as follows: government purchase and control, broad scope of implementation, diverse purposes, unified payment methods and standards and low efficiency. Examples are the Conservation Reserve Program (CRP) of USA, Hydrological Environmental Service Payment Program (PSAH) and Environmentally Sensitive Area (ESA) and Countryside Stewardship Scheme(CSS) of UK. The second model is a market-oriented one including green reimbursement, quota trading, eco-labelling systems, emission license transaction and international carbon trading, etc. The main features of this model are payment by users or beneficiaries of ecological services, voluntariness, management by intermediaries, a small range of implementation, singular purposes, diverse methods of compensation and high efficiency. For example, PROFAFOR carbon sequestration project in Ecuador, Los Negros drainage basin conservation project in Bolivian and Vittel water conservation project in France fall into this category. There practices of ecological payment have achieved certain success because of sound laws and regulations, strong government payment capability, well-established property right systems, mature market mechanisms and participation of different stakeholders, etc.

Starting from 1970s and 1980s, ecological compensations began to be carried out in Sichuan, Yunnan and other provinces in China. Since middle and late 1990s, pilot projects have been implemented in forests, grasslands, mines and many sectors and ecological compensation have been advanced rapidly. Three provinces of Sichuan, Shaanxi and Gansu took the lead in starting pilot projects of returning farmland to forestry, which were later expanded to over 20 provinces in 2002. Guangdong, Fujian and other provinces carried out pilot project of river basin ecological compensation. In the last few years, great improvement has been made regarding the depth and breadth of the pilot projects. Shenzhen Municipality began to conduct ecological compensation in Dapeng Peninsula in 2007. A special subsidy of ecological conservation was distributed to 17,000 villagers (each villager received 1,000 RMB per month) in the peninsula through financial transfer payment, which stimulated the enthusiasm of the villagers and improved the ecological environment of Dapeng New Area. Coordinated by Ministry of Finance and Ministry of Environmental Protection, Zhejiang and Anhui kicked off a three-year cross-province river basin ecological compensation project in Xinanjiang River Basin in 2011, the first project of its kind in China. The central government and the two provinces jointly made the investments and determined the allocation of funds according to water quality assessment results. The water quality in Xinanjiang River Basin was rated excellent in general in the three years and met the requirement set by the project. In recent years, cross-provincial watershed ecological compensation in Jiuzhou River, Tingjiang – Hanjiang, Dongjiang and other rivers has gradually been advanced. These practices contribute to win-win and equitable development of different regions.

Restrictive Factors for Ecological Compensation

Although China has accumulated rich experience in ecological compensation in the fields of forest, grassland, wetland, river basin and other key ecological functional areas and achieved the preliminary results, lack of special laws, insufficient compensation, insufficient supporting systems and other problems are becoming more and more serious, restricting the advancement of ecological compensation.

Firstly, there is no special law regarding ecological compensation. The existing regulations are too fragmented and in principle. No. 1, special legislation on ecological compensation is missing. Although *Ecological Compensation Act* was listed into legislative plans in 2010, it has not been promulgated because ecological compensation involves complex interests and core contents such as compensation principles, fields, bodies, capital, methods, targets and performance evaluation, etc. are still under discussion, which has made it very difficult to make effective and comprehensive ecological compensations. No. 2, the existing regulations on ecological conservation are not so effective because they are too fragmented, unsystematic and in principle. *Law on Environmental Protection, Grassland Law, Forest Law and Mining Resources Law*, etc. have poor operability because they just lay down some general principles on ecological compensation. Local regulations and rules mainly concern forest, mining and other fields. They just have narrow coverage and low legal effects and thus fail to provide legal basis for ecological compensation.

Secondly, problems regarding ecological compensation such as narrow scope, low standard, limited methods, inadequate sources of funding and lack of supervision still exist. No.1, existing ecological compensation mainly concentrates on forestry, grassland, mining resource development and other fields. Ecological compensation on river basins, wetland, ocean, land and desert, etc. is still in its infancy. Atmosphere and other fields have not been included. No. 2, the current compensation standards are generally very low and cannot cover cost of ecological conservation and opportunities cost of development, let alone service value of ecological systems. No. 3, methods of ecological compensation are limited. At present, ecological compensation mainly takes the form of financial subsidies. Methods related to policy, materials, technology and talent such as industrial support, technical assistance, HR support, employment and training have not been duly recognized and promoted. No. 4, sources of funding for ecological compensation are limited and supervision is inadequate. At present, funding is mainly from transfer payment by the central finances. The horizontal ecological compensation between regions, between downstream and upstream and between different social groups is not insufficient. Market transaction mechanisms for ecological products are not sound. In some areas, ecological compensation funds are not properly used and some funds have been squeezed or misappropriated.

Thirdly, supporting mechanisms such as property right system, surveillance systems and calculation methods need to be improved. No. 1, the property right system is not sound. Well-defined property right system is the prerequisite for the distribution of benefits, the determination of compensation and confirmation of rights and responsibilities. However, the property right of natural assets in China is not clarified. The immature property right system has led to the predatory use of resources. The rights and responsibilities of the beneficiaries, protectors, polluters and victims are not clear, which has made it very difficult to achieve a fair and effective distribution of ecological benefits. No. 2, the surveillance systems are not perfect with disconnected surveillance functions, fragmented surveillance networks, diverse surveillance methods, different surveillance indicators and inadequate capacity building for surveillance staff, technologies and funding. No. 3, the calculation methods are not unified. There has been no consensus regarding methods used to dertermine ecological standards. Currently, opportunity cost method, intention survey method and market method are being employed. Also there have been controversies regarding diverse methods on calculating ecological service value which include shadow engineering method, equivalent method and market value method, etc.

It Is Imperative to Promote Ecological Compensation

The essence of ecological compensation is to re-allocate the development fruits and re-cut the resource cake. It is an effective way to promote the economic transformation, social equity and ecological optimization and coordination. We must unswervingly promote ecological compensation.

Firstly, legislation is the fundamental guarantee of ecological compensation standardization. No. 1, Regulation on Ecological Compensation, its implementation measures and technical guideline should be issued as soon as possible. In the top level design, the State should clarify the principles, fields, targets, standards, processes, sources of funding, assessment mechanisms, accountability, rights and obligations of stakeholders, etc., which can establish a complete legal framework and provide an institutional guarantee for the standardized operation of ecological compensation. No. 2, relevant laws and regulations need to be amended so as to be harmonized with the special law on ecological compensation and build a multi-level and comprehensive ecological compensation mechanism covering water, soil, forest, grassland, sea, desert and wetland, etc.

Secondly, compensation standards, scope, sources of funding and diversified methods are the key factors to fully implementing ecological compensation. No.1, a dynamic and systematic ecological compensation standard should be developed. Different standards covering grassland, forest, mining resources, desert and river basins and key ecological functional areas should be formulated respectively so as to develop a standard system. During implementation, ecological compensation standards should be adjusted in line with conservation effect, government financial strength, social economy and other factors. No. 2, we should expand compensation scope, broaden sources of funding and enrich compensation methods to build a diversified ecological compensation mechanism. Built on the experience gained from the pilot projects in forest, mining resources, grassland, river basin and other fields, we can expand the scope to cover atmosphere, wetland, key ecological functional areas in order to achieve an integrated conservation of the ecological system. We should enhance financial support, optimize financial expenditure structure, improve the market-oriented compensation mechanisms including transactions of emission permits, water rights and carbon sink, etc., encourage and guide the participation of private capital and gradually set up a diversified investment and financing mechanism with government guidance, market promotion and private involvement. We should encourage benefiting areas to assist conservation areas with diversified horizontal compensation methods including industrial collaboration, industry transfer, talent training, jointly building industrial parks, etc. in addition to capital compensation. The diversified compensation methods can help to enhance the functions of "blood replacement" and "blood generation" in the compensation areas. These poor areas can get financial compensation to develop ecological economy, build infrastructure, train talents, protect the environment and shake of poverty.

Thirdly, perfecting the property right system, improving the surveillance system and enhancing technical support are powerful guarantee for orderly, accurate and efficient implementation of ecological compensation. No. 1, the property right system should be perfected. Ecological compensation not only adopts administrative mechanism, but also market-oriented methods. The prerequisite to optimizing public resources is to clarify property rights. The property right system with balanced rights and obligations of the main players, a well-defined boundary of property rights, a complete ownership and equitable distribution of interests are conducive to implementing ecological conservation. No. 2, the ecological surveillance system should be improved. A sound surveillance system can standardize monitoring methods, improving indicators, establish a unified surveillance network with smart and standardized surveillance technologies, achieving synergies and provide complete, coherent and dynamic surveillance information. No. 3, the technical support system should be strengthened. Geographic information technology research and application should be enhanced. We can identify changes in resource utilization with automatic remote sensing tools, develop inventories of ecological assets, build a balance sheet of natural assets and achieve lean management by linking per-unit-area value with ecological compensation. We need to unify ecological standards and clarify the method of calculating ecological service value. We should actively explore the method of building a quantitative assessment system for ecological compensation.

Fourthly, enhancing process supervision, improving performance evaluation and mobilizing social participation are driving forces for ecological compensation. No. 1, supervision and accountability should be strengthened. An coordination mechanism of ecological compensation should be established to quantify the powers and responsibilities of various departments and enhance supervision and inspection of the allocation and use of funds in key fields and regions. Those departments with poor performance must be held accountable. No. 2, performance evaluation should be improved. It should be made clear that dynamic assessment on ecological compensation will be held regularly. Regions with good performance should be given policy preference and financial incentives. No.3, social participation should be mobilized. We must enhance publicity and education on ecological compensation, develop an open and fair compensation procedure, give full play to the role of the media supervision, raise public awareness and support for ecological compensation, guide all kinds of beneficiaries to fulfill their ecological compensation obligations and urge regulators to fulfill their ecological governance responsibility.

第十二章 技术创新引领绿色发展新动力

迄今为止，人类社会发展大致经历了原始文明、农业文明和工业文明三个阶段，开始向生态文明时代迈进。原始文明时期，人类依附于自然、受自然支配；农业文明时期，人类转而利用、改造自然，但对自然的扰动并不剧烈；到了工业文明时期，人类控制自然、消耗自然、严重破坏自然环境，并由此引发了一系列生态问题。而在生态文明时期，人类致力于创造既生态——人与自然和谐互动，又文明——人类改造、利用自然的方法和能力大幅提升的文明形态。历史的经验告诉我们，每一种文明形态都有其对应的主流技术形式。文明形态与技术发展之间是相互推动、相互促进、相互影响的辩证关系。20世纪70年代以来，绿色技术逐渐进入人们的视野。《21世纪议程》等文件指出绿色技术是获得持续发展，支撑世界经济，保护环境，减少贫穷和人类痛苦的技术。充分认识绿色技术，促进绿色技术的推广与有效应用是实现生态文明的关键。

绿色技术是多维度多领域的革命性创新

绿色技术涵盖了多维度、多领域的技术创新，它将"环境友好"作为创新过程的重点，致力于在绿色环保的目标导向下实现技术选择（如风电与煤电）与技术完善（如污染处理）。绿色技术不是单一的技术，而是复杂的技术群，覆盖可再生能源、节能环保材料、清洁生产工艺、污染治理、环境监测、生态修复等诸多领域，被广泛应用于生产、生活的方方面面。

世界各国为实现发展转型升级，纷纷以绿色技术为突破口，寻求建树。英国的绿色建筑发展历史悠久、推广度高、实践丰富，是绿色建筑的行业标杆；日本提出"绿色发展战略"总体规划，突出开发可再生能源和节能主题的新型机械，计划将大型蓄电池、新型环保汽车和海洋风力发电建设成为日本绿色增长战略的三大支柱产业；巴西利用其广袤的耕地和发达的农业，重点发展生物能源和新能源，并以此推动新能源汽车发展；印度则重点关注太阳能资源的推广。

历经四十余年的自主发展和技术引进，我国已基本形成涵盖从源头（工业生产、居民生活）到末端（污染治理、生态修复）的绿色技术体系，包括能源技术、材料技术、催化剂技术、分离技术、生物技术、资源回收及利用技术等。其中，太阳能光伏发电、新能源汽车、水污染处理及水体修复等领域的技术处于世界先进水平。然而，总体上看，我国绿色技术在生产、生活环节的应用与推广仍相对迟缓。例如，2003年《清洁生产促进法》实施以来，工业企业清洁生产技术改造方案实施率仅为44.3%。此外，我国的绿色建筑起步较晚，从1986年颁布北方居住建筑节能设计标准开始，只有北京、上海、天津等少数大城市有较理想的节能建筑，能源利用效率仅为33%，与发达国家相差10个百分点。全国共有3000余项目获得绿色建筑评价标识，而其中近95%的项目只是取得了绿色建筑设计标识，在竣工后获得运行标识的仅占总数的5.8%。生物柴油等绿色能源由于扶持推广政策力度不足，同时缺乏强制性措施和配套发展机制，产业发展举步维艰。我国光伏制造业虽发展迅速，但对国外市场的依赖反映出国内绿色应用的巨大缺口。新能源汽车发展如火如荼，但更大范围的绿色交通系统的建立受制于固化的交通现状，因而存在巨大困难。

绿色技术创新推广任重道远

绿色技术重要的战略地位、先进的技术性能以及巨大的需求缺口与其艰难推广的现状存在强烈反差。同其他技术一样，绿色技术内嵌于经济社会系统之中，新技术的采用，必须顺应经济社会系统发展与变革的潮流，并能反作用于经济社会系统，重塑生产、生活。必须深入分析、深刻认识绿色技术推广存在的困难。

一是技术进化的长期性决定了绿色技术不能一蹴而就。纵观技术发展的历史，新技术的应用从来不能够立竿见影；相反，每一种新技术都需要经受经济社会系统发展的长期历练。蒸汽船具有其巨大的效率优势，彻底取代帆船尚且用了一个半世纪的时间；汽油汽车从因石油的偶然发现加入竞争行列，到最终确立主导地位，历时百余年之久。电动汽车主要优势为绿色环保，但相较内燃机汽车尚显动力不足，因此电动汽车要替代对传统的内燃机汽车也很可能需要较长的时间。

绿色技术发展至今，长则三四十年，短则不过数月，大多处于"胚胎"时期。由研发到可推广阶段，再到技术的广泛应用，直至各类配套设施的跟进，绿色技术还需要漫长的转化期，推广绿色技术任重而道远。

二是原有主流技术的竞争优势决定了绿色技术推广的艰难性。从经济的角度来看，主流技术背后必然有与之长期适应的生产设备与工艺、制度乃至消费者习惯，由此形成巨大的比较优势。这些因素在很大程度上削弱了绿色技术的竞争优势，使其短期内难以成为主流。例如2015年并网光伏发电的标杆上网电价为燃煤发电的两倍以上，火力发电仍占据大量市场。甘肃、新疆等西部省区的光伏发电，由于消纳能力有限，造成"弃光"现象。再如电动汽车，由于充电设施与油站、气站相比，分布稀疏不均，严重制约了长途运输以及市区交通以外的出行需求，因此电动汽车也难以在短时间内取代具有完善配套设施的燃油汽车。

三是经济社会系统的短视性造成绿色技术创新推广动力不足。绿色技术的应用和推广归根结底是为社会服务的。然而，相对于发达国家，我国还缺乏不遗余力地支持绿色技术发展的勇气，由此导致绿色技术产品市场热度低，内需疲软。以德国为例，为发展可再生能源，政府对传统能源征收最高达47%的可再生能源附加费，使可再生能源电价低于传统能源电价，有力地支持了光伏发电的发展。此外，商品种类稀少、价格高昂等原因使绿色技术的接纳度还不够高。如低能耗的供暖技术、节能家用电器等的应用未得到广泛认可，导致建筑能耗整体偏高。据统计，2014年，我国建筑能耗占社会总能耗的45%以上。总的来说，政府、企业和消费者往往着眼短期利益，没有长远目标和规划，创新推广绿色技术动力不足。

推广绿色技术要找准发力点

绿色技术与经济社会系统紧密相连，推广应用绿色技术不应仅着眼于技术自身，还需要统筹经济社会的发展和变革。

首先，实施利基战略，为创新的绿色技术提供保护和发展机遇。利基是指在创新技术不具有市场竞争优势的起步阶段，为其创造一个独立于正常竞争市场的生存和发展的空间以避免其在市场竞争环境下夭折。目前，对创新的绿色技术的扶持呈现多元化的特点，既包括政府扶持，如税收减免、补贴、改善融资环境、简化行政审批程序等，也包括社会力量的扶持，如社会资本进入创新绿色技术的应用领域。通过对创新绿色技术的扶持，可以帮助这些技术在细分市场取得竞争优势。如光伏技术在自发自用的分布式光伏应用市场，无土栽培技术在山东寿光的盐碱地中皆取得了竞争优势。在细分市场上积累竞争优势，可以为创新绿色技术最终取得对主流技术的竞争优势打下良好的基础。

其次，推动技术创新，提高采用绿色技术企业的自生能力。自生能力是指在自由竞争的市场经济中，一个正常经营的企业在没有外部扶持的条件下，能够获得不低于社会可接受的正常利润水平的能力。正所谓"打铁还得自身硬"，如果采用绿色技术的企业不具有自生能力，再强的扶持力度也是枉然。提高自生能力的关键在于增加企业的经济收益，这要通过创新提升企业生产环节的全要素生产率来实现。以绿色建筑为例，目前的绿色建筑材料成本高昂，造成绿色建筑售价高企，普通群众基本无人问津。因此迫切需要推动技术创新，降低绿色技术企业的成本，提升企业的自生能力。

最后，改善激励机制与舆论氛围，刺激绿色技术的应用需求。绿色技术的应用和推广在启动期需要政府的大力引导，但根本上需要市场需求的支撑。由于缺乏比较优势，当前绿色技术产品需求疲软。绿色技术有明显的正外部性，政府部门有责任在技术选择和技术完善方面给予激励，通过技术标准、排污税费、补贴等政策工具营造更加有利于绿色创新的有利环境。例如加大财政资金中科技支出的比重、加大绿色产品的政府采购比重、促进绿色技术项目示范等。同时，必须探索建立将资源环境外部性内部化的机制，通过征收资源环境税，提振对绿色产品的需求。从长远来看，要通过促进公众参与和监督绿色发展政策的制定与实施，加大生态文明建设理念的宣传力度，引导群众牢固树立生态环境忧患意识，最终营造绿色发展、绿色消费的社会氛围。

Chapter 12
Technological Innovation: A New Driving Force for Green Development

So far, the development of human society has gone through three stages: primitive civilization, agricultural civilization and industrial civilization and began to move towards the era of ecological civilization. During the first stage, human beings are subordinated to nature and dominated by nature. During the second stage, mankind began to take advantage of and transform nature, but the disturbance to nature was not dramatic. In the period of industrial civilization, human beings began to control and consume nature and thus caused serious destruction of the natural environment, triggering a series of ecological problems. In the period of ecological civilization, mankind is committed to creating an ecological world characterized by harmonious interaction between man and nature and a civilized society characterized by a significant improvement in methods and ability to transform nature. The experience of history tells us that every form of civilization has its corresponding mainstream technical form. The relationship between civilized forms and technological development is a dialectical one characterized by mutual promotion and mutual influence. Since 1970s, green technology has gradually entered people's vision. *Agenda 21* and other documents point out that green technologies can achieve sustainable development, support the world economy, protect the environment and reduce poverty and human suffering. Fully understanding green technologies and promoting the effective application is the key to the realization of ecological civilization.

Green Technology Present a Multi-dimensional and Multi-field Revolutionary Innovation

Green technology covers multi-dimensional and multi-disciplinary technology innovation, which regards "environmentally friendly" as the focus of the innovation process. Guided by green development and environmental protection, green technology is committed to the goal of achieving technical options (such as wind power and coal) and technical improvements (such as pollution treatment). Green technology is not a single technology, but a complex technical set covering renewable energy, energy saving and environmental protection materials, clean production processes, pollution control, environmental monitoring, ecological restoration and many other fields. Green technology is widely used in all aspects of human production and life.

Countries in the world wish to use green technology as a breakthrough point in order to achieve green development and economic transformation. Regarding green building technologies, UK is the leading country in the world with a long history, rich experience and wide application. Japan put forward the master plan of green development strategy, highlighting the development of new machineries using renewable energy and energy saving technologies and regarding large storage batteries, new environmental-friendly vehicles and maritime wind power as the three pillar industries for Japan's green development strategy. Taking advantage of its vast land and developed agriculture, Brazil focuses on developing bio-energy and new energy which can be used to promote the development of new energy vehicles. India attaches great importance on the application of solar energy resources.

After more than 40 years of independent development and technology introduction, China has basically established a green technical system covering the whole process from the source of pollution (industrial production and residents' life) to end (pollution control and ecological restoration), including energy technology, material technology, catalyst technology, separation technology, biotechnology, resource recovery and utilization technologies. Among them, solar photovoltaic power generation, new energy vehicles, water pollution treatment and water remediation and other areas of technology are leading the world. However, on the whole, China has been quite slow in applying and promoting green technologies in production and people's daily life. For example, since the implementation of the *Cleaner Production Promotion Act in 2003*, the implementation rate of technological transformation projects for cleaner production in industrial enterprises was only 44.3%. Furthermore, green buildings started very late in China. Since the promulgation of design standards on energy conservation in residential buildings in North China in 1986, only a small number of big cities such as Beijing, Shanghai and Tianjin, etc. have built energy saving buildings with energy efficiency reaching only 33%, a 10% lower than that in developed countries. A total number of 3,000 projects in China obtained green building evaluation marks. However, 95% of the projects only obtained green building design logos. Only 5.8% of projects obtained green operation logos after completion. Development of green energies like bio-diesel is in a difficult situation due to lack of supporting policies and compulsory measures. Although China's PV manufacturing industry is developing rapidly, high dependence on foreign market reflects a big gap in domestic green applications. The development of new energy vehicles is in full swing, but the establishment of a wider range of green transport system encounters great difficulty because of some inherent problems.

Innovating and Promoting Green Technologies Still Have a Long Way to Go

There has been a strong contrast between the important strategic position, advanced technical performance and huge demand gap of green technologies and the current situation of considerable difficulty in promotion. Like other technologies, green technology is embedded in the economic and social systems. The adoption of new technologies must follow the trend of development and change of these systems. New technologies can also react upon the economic and social systems by reshaping production and life. We must conduct in-depth analysis and have a profound understanding of the difficulties in promoting green technologies.

Firstly, the long-term evolution of technology determines that green technology cannot be achieved overnight. Throughout the history of technological development, the application of new technology can never get instant results; on the contrary, every new technology needs to withstand the test of the development of the economic and social systems. It took steamboats half a century to replace sailboats although the former had huge advantages in efficiency over the latter. It also took gasoline automobiles over 100 years to become dominant in the transportation industry. The main advantages of electric vehicles are green and environment-friendly, but vehicles with internal combustion engines are more powerful. Therefore, it may take a long time for the former to replace the latter.

Most green technologies are still in their infancy because some of them were developed 30 or 40 years ago and some were developed months ago. From R&D to stages of technical promotion, wide application and the supporting facilities becoming available, green technologies need a long conversion period, so promoting green technologies still has a long way to go.

Secondly, the competitive advantages of the existing mainstream technologies make it very difficult to promote green technologies. From an economic point of view, behind the mainstream technologies there must be production equipment, processes and consumer habits which have been supportive to these technologies, creating a huge comparative advantage. These factors largely weaken the competitive advantages of green technology, making it difficult for green technology to become mainstream in the short term. For example, the price of electricity generated by PV is over twice as high as that of electricity generated by coal-fired power plants. Thermal power still occupies a large market share. PV power generated in Gansu, Xinjiang and other western provinces and autonomous regions has to be abandoned due to limited utilization capacity. Another example is electric vehicles. Compared with gasoline stations and gas stations, charging facilities are not evenly distributed, which has severely restricted long-distance transportation and out-of-town transportation needs, so it is difficult for electric vehicles to replace fossil fuel vehicles within a short time.

Thirdly, there has been lack of motivation in promoting green technology due to short-sightedness of economic and social systems. The application and promotion of green technology eventually serve the society. However, compared with developed countries, China lacks the courage to fully support the development of green technology, which leads to a sluggish market and weak demand for products related to green technology. For example, Germany collects at most 47% renewable energy surcharge over traditional energies in order to develop renewable energy so that price of electricity generated by renewable energies is lower than that by traditional energies, which has vigorously supported the development of PV power generation. In addition, green technology is not well received due to limited number of products and high price. For example, heating technology with low energy consumption and energy-saving household appliances, etc. are not widely recognized, resulting in high energy consumption in buildings. According to statistics, building energy consumption accounted for over 45% of the total energy consumption in China in 2014. In all, governments, enterprises and consumers usually focus on short-term interests and lack long-term goals and planning, which causes inadequate motivation in promoting green technology.

Identifying the Focal Points to Promote Green Technologies

Green technology is closely related to social and economic systems. In order to promote the application of green technology, we should not only focus on technology itself, but also coordinate economic and social development and change.

Firstly, we should implement niche strategies to provide protection and development opportunities for innovative green technologies. Niche means the creation of a space of survival and development independent of normal market competition at the initial stage when innovative technologies do not have competitive advantages to protect new technologies from dying a premature death in the market competition. At present, there are diversified support for green technology including both government support, such as tax deduction and exemption, subsidy, improving financing environment and simplifying administrative review and approval, etc. and support from the private sector, e.g., private capital entering application fields of innovative and green technologies. These supports can help the green technologies to gain competitive advantages in the market segment. For example, PV technologies in the distributive PV application market and soilless culture technology applied in the saline-alkaline soil in Shouguang City, Shandong Province have both gained competitive advantages. Innovative green technologies need to accumulate comparative advantages in the segmented market, which can lay a good foundation for green technologies to gain competitive advantages over mainstream technologies.

Secondly, we should enhance the self-sustaining ability of enterprises adopting green technologies in order to promote technical innovation. Self-sustaining ability means that in a market economy with free competition, an enterprise operating normally, without any external support, can make normal profit which is no less than socially acceptable. Just as the saying goes, to forge iron, one must be strong. If the enterprises adopting green technologies do not have self-sustaining abilities, even strong support will be in vain. An enterprise has to improve its ability to generate more profits, which can be achieved through increasing total factor productivity in the production link via technical innovation. Take green buildings as an example. At present the cost of green building materials is too high, resulting in high sale price of green buildings and the general public cannot afford them. Therefore, we need to promote technical innovation to reduce the cost of green enterprises and enhance their self-sustaining abilities.

Last but not least, we should improve the incentive mechanism and public opinion to stimulate the needs for the application of green technologies. At the initial stage of application and promotion of green technologies, government guidance is necessary, but market demand is fundamental. However, the demands for products of green technologies is weak at present due to lack of comparative advantages. Green technologies have obvious positive externality. Government departments have the responsibility to incentivize technical selection and improvement and create favorable environment for green innovation through technical standards, pollution discharge taxes or fees, subsidies and other police tools. The government can earmark more fiscal funds for science and technology, increase the proportion of green products in government procurement and design more demo projects to promote green technologies, etc. Meanwhile, we must establish mechanisms to internalize the externality of resource environment and boost the demand for green products through collecting resource and environment taxes. We should vigorously publicize the concept on ecological civilization construction through promoting public participation and supervision in the development and implementation of polices on green development and raise public awareness on ecological environment in order to create a social atmosphere of green development and green consumption.

第十三章 智能技术助力生态治理现代化

21世纪以来，以移动互联网、云计算、大数据等为代表的智能化、信息化技术为世界经济发展注入了新的动力，从根本上改变了传统的生产方式和消费习惯。无处不在的"慧眼"能够解决环境信息破碎化、局地化问题，使监测、预警更加准确高效，有效打破公众参与的信息壁垒。智能技术是产业转型升级的有力引擎，在加速淘汰低端产业的同时引导企业向绿色化生产转型，已成为推进生态文明建设领域治理体系和治理能力现代化的重要手段。

既是"紧箍咒"又是"催化剂"：智能技术推动产业升级与绿色转型

智能技术不断推进产业升级与绿色转型，成为可持续发展的强劲动力，对经济社会生态系统的全局长远发展产生重大的引领带动作用。

第一，加强监控力度，促进企业绿色清洁生产。对于企业而言，智能技术的广泛应用，意味着违法排污将被更多双眼睛盯着、更多"紧箍咒"压着，排污所付出的代价将会越来越重。以往污染源在线监控的过程中，千奇百怪的造假术层出不穷：设法将污水探头探到的废水稀释，或在排放高浓度大剂量的废水时，将探头挪放到另一个位置等，造成时空错位、监控失灵。智能技术的广度、深度使得造假越发困难，大数据、云计算等先进信息化技术交叉印证，使弄虚作假在多维度的智能技术的视角中无处遁形。通过搭载大数据模型和物联网技术，能实时掌握监管区域内各排放企业的产污、治污、排污情况，并对企业的排污行为进行分析，及时发现违规行为并发出预警，确定环保重点监管对象。互联网信息的公开和透明化，让资源环境压力大的低端产业捉襟见肘。

第二，智能技术助推绿色转型。随着计算能力的增加和计算成本的降低，智能技术逐渐应用并深刻推动着绿色发展的进程。大数据、云计算、物联网等智能技术的普及,为产业的智慧管理、智慧交换、智慧共享提供了平台,在淘汰落后产能的同时,带动基础设施建设科学化以及产业选择与发展方向合理化。同时,智能技术的应用使产业将用户端的价值需求作为整个产业链的出发点,改变以往的工业价值链从生产端向消费端、从上游向下游推动的模式,各个环节的配合更加优化。智能技术以数据为核心,对接供给和需求,让产业变得智慧,成为生态文明背景下产业升级的催化剂。

近年来,逆向物流回收技术不断应用于环保领域,为降低处理回收物品的成本,推动资源再生利用开拓了方向。在生活垃圾回收方面,杭州部分城区开始试点实行"一户一码",居民在垃圾袋上须粘贴二维码贴纸,在垃圾房前通过扫描二维码才能打开相应的房门进行垃圾投放。在电子电器回收方面,"E环365"通过搭建废旧电器电子产品回收服务信息平台,为社会提供盘活消费者废弃电器电子资源,促进再生资源综合利用的信息服务,引导大家自觉将废旧电子产品送到正规渠道回收利用。总之,当下"物联网""人工智能服务"技术在"互联网"的基础上,将其用户端延伸和扩展到任何物品与物品之间,通过人与人、人与物、物与物的相连,进行信息交换,从而解决信息化的智能管理和决策控制问题。

既是"千里眼"又是"智囊团":智能技术助力精准监测与科学决策

智能技术通过实时、海量监测数据和准确的预测预警,促进政府管理的智能高效化和决策的科学理性化,改变了长期以来依靠经验、理论和思想的决策方式,直觉判断让位于精准的数据测控、分析。

第一,智能技术助力精准监测、预报预警。随着对环境监测、预警重视程度的不断提高,我国已形成了国家、省、市、县4级环境监测网络,目前已有国控的空气质量监测网站1400多个。海洋生态环境监测工作开展了管辖海域海水质量、沉积物质量、生物多样性状况趋势监测,全方位跟踪海洋生态变化动态。智能技术的动态综合运用,显著提高了环境预报预警的准度和精度,使得环境监测网的覆盖范围不断扩大,监测内容更加丰富细致,监测能力与预警能力都大幅提升。在季风作用下，新加坡的烟霾常常受到其他地区的影响，为此新加坡牵头开发了东南亚国家联盟区域烟霾预警系统。这一预警系统基于卫星成像等智能技术，能够及时发现林火并准确定位，为预测烟霾发生提供了科学依据。

第二，环保大数据支撑科学决策。政府部门可借助互联网服务平台，采集环境监测数据，并与社会公众需求信息、民意诉求信息等相结合，形成环保大数据。通过对环保大数据的分析，能够揭示数据之间的关联，发现现象背后的规律，提高生态环境治理的精准性和有效性。在大数据的外部数据获取与信息基础设施建设上，我国大数据在数据量、多样性、时效性和价值密度等方面均已较为完善，但数据的开放性、共享性及数据处理的创新性和科学性还都有待提高。通过进一步完善大数据分析的技术和人才队伍，政府部门能够及时了解污染物排放状况，精确评估环境质量及变化趋势，准确预测、预警各类环境污染事故的发生概率，提高环境形势分析能力。环保大数据是提高我国环境管理信息化水平的重要手段，利用大数据技术能够对环境管理中看似毫无关联的碎片化信息进行关联分析，从中发现趋势、找准问题、把握规律，实现"用数据决策"，有助于增强生态治理的精准性和有效性，提高政府管理决策的水平。

既是"万花筒"又是"传声筒"：智能技术确保信息公开与公众参与

公众的参与是生态治理的有力保障。智能技术的发展有利于公众及时获取环境信息，为全民参与生态治理决策、监督生态治理效果提供了便利。

第一，打造多元化平台，增强环境信息可得性。环境与公众的切身利益息息相关，以往公众不是不愿意参与生态治理，而是受到环境信息条件的限制无法有效参与。环境信息披露的全面性和真实性常常难以保证。目前，国家重点监控的企业都有在线监测设备与环保部门联网，而且环保部门还会进行每季度一次的监督性监测并公布。从2015年起，全国338个地级及以上城市空气质量监测数据实现了实时发布全覆盖，所有数据在环保部的网站上均可查询。一些社会环保组织也开始使用移动互联网技术推进环境信息的公开，如北京公众环境研究中心开发的"蔚蓝地图"，通过 APP、网站等途径，除了展示全国各地公布的空气、江河质量，还包括企业最新排放数据。通过移动互联网应用软件随时查询 PM2.5 浓度等空气质量，已成为很多老百姓的生活习惯。环境信息的公开使得公众可以随时随地获取环境数据，为公众的知情参与提供了必要支持。

第二，创设多样化渠道，激发公众参与热情。新媒体的兴起，特别是微博、微信等新兴社交媒体的普及，使公众监督权利的行使更为灵活和便利，为公众参与生态治理提供了更即时的渠道和更广阔的平台。"人人都是观察员，人人都是监督员，人人都是环保员"的理念逐渐成为现实。各种环保微公益活动在社会化新媒体中得以广泛尝试，例如"随手拍家乡污染""随手拍黑烟囱"等活动在微博中成为热门话题，唤醒了公众的监督意识，使公众参与生态治理的热情高涨。政府部门可以将公众上传的碎片化数据进行分析和整合处理。此外，社交媒体上公开的海量数据，还可帮助政府了解公众需求，进而提供差异化和精细化的公共服务，改善公众对于生态治理的认知和体验。

Chapter 13
Smart Technologies Promote Modernization of Ecological Governance

Since the beginning of the 21st Century, smart technologies on IT represented by mobile internet, cloud computing and big data, etc. have injected new momentum into the global economic development, which has fundamentally changed traditional production modes and consumption habits. The ubiquitous "smart eyes" can resolve issues of fragmented and localized information, make surveillance and early warning more accurate and efficient and break down information barriers for public participation. As powerful engines for industrial transformation and upgrading, smart technologies can guide enterprises to green transformation while accelerating the elimination of low-end industries, thus becoming an important means to promote eco-civilization governance system and modernization of governance capability.

Smart Technologies Are Both "Inhibitors" and "Catalysts" for Promoting Industrial Upgrading and Green Transformation

As a strong driving force for sustainable development, smart technologies continuously promote industrial upgrading and green transformation and play an important role in the long-term holistic economic, social and ecological development.

Firstly, smart technologies can force companies to produce in a clean and green way as a surveillance tool. The application of smart technologies means that companies who discharge pollutants illegally will be watched by more eyes and be inhibited from discharging at random, so the cost of discharging pollutants will be increasingly high. In the past, polluters used various methods to avoid being detected by online surveillance tools when discharging pollutants. For example, they tried to dilute waste water detected by waste water detectors, or they moved the detectors to another place. Smart technologies make it more and more difficult to tamper with data and advanced information technologies as big data and cloud computing can prevent companies from falsifying data. Regulators can monitor in real time companies regarding their production and pollutant discharge and control through the use of big data models and internet of things and analyze the discharging behaviors of companies, issue early warnings and identify key supervision targets in environmental protection. The openness and transparency of internet information will force low-end enterprises under tremendous environmental pressure to close down.

Secondly, smart technologies help to boost green transformation. With the enhancement of computing power and decrease of computing cost, smart technologies began to be applied gradually and have considerably promoted green development. The popularization of smart technologies such as big data, cloud computing, internet of things (IOT) etc. provides a platform for smart industrial management, exchanges and sharing. While phasing out the backward production capacities, they can drive the construction of infrastructure, industrial selections and development directions in a scientific and reasonable way. Meanwhile, with the application of smart technologies, the value demands of the users have become the starting point of the whole industrial chain, which has changed the previous model of industrial chain pushing from production side to consumption side and from upstream to downstream. All the links in the industrial chain have become more optimized. Smart technologies with data at the core linking supply and demand enable the industries to be more intelligent and thus become the catalysts for industrial upgrading in the context of eco-civilization.

In recent years, reverse logistics recycling technology has been continuously applied in the field of environmental protection, which has opened up a new direction for decreasing recycling cost and promoting resource recycling. Regarding domestic garbage collection, a pilot program of "one household, one code" began to be implemented in Hangzhou, east China's Zhejiang Province. Residents are required to adhere stickers with QR codes to garbage bags and they need to scan the codes so as to open corresponding doors to dispose of the garbage. Regarding recycling of electronic and electrical products, E-ring 36 establishes an information platform for recycling the waste products, providing information services to promote comprehensive utilization of resources and encouraging everyone to send waste products to regular recycling channels. In all, on the basis of internet technologies, technologies such as IOT and AI help to conduct exchanges of information between people and people, between people and things and between things and things through extending and expanding the recycling chains so as to resolve issues like information-related smart management, decision making and control.

Smart Technologies Are Both "Telescopes" and "Think-tanks" for Accurate Surveillance and Scientific Decision Making

Through real-time and massive surveillance data and accurate forecast and early warning, smart technologies can promote intelligent and efficient government management and scientific decision making and have changed the long-term decision-making modes of relying on experience, theory and ideas. Therefore, accurate data monitoring and analysis has replaced intuitive judgement.

Firstly, smart technologies boost accurate surveillance, forecast and early warning. With increasing attention to environmental surveillance and early warning, China has set up a four-tier environmental surveillance network at the levels of state, province, city and county. So far over 1,400 air quality surveillance sites have been established. Regarding maritime eco-environment surveillance, surveillance of sea water quality, sediment quality, bio-diversity trends have been carried out so as to comprehensively monitor maritime ecological changes. Dynamic application of smart technologies has remarkably improved the accuracy of environmental forecast and early warning, expanded the coverage of environmental surveillance network, enriched the surveillance content and enhanced the capability of surveillance and early warning. Smaze in Singapore is often affected by other regions due to monsoon, so Singapore plays a leading role in jointly developing a smaze early warning system for ASIAN region. Based on satellite imaging and other smart technologies, this early warning system can detect forest fires in time and conduct accurate positioning, which can provide scientific evidence on the forecast of smaze.

Secondly, environmental protection data support scientific decision-making. Government departments can use the internet service platforms to collect environmental monitoring data which can constitute environmental protection big data in combination with information on public demand and public opinion, etc. Through the analysis of large environmental data, people can reveal the relationship between the data, find the laws behind the phenomenon and improve the accuracy and effectiveness of ecological and environmental governance. Regarding external data acquisition and information infrastructure construction of large data, China has made significant improvement on data volume, diversity, timeliness and value density and other aspects, but there is still great room for improvement regarding data openness, sharing and innovative and scientific data processing. By further improving the technology and talent pool on large data analysis, government departments can keep abreast of the situation on pollutant discharge, accurately assess the environmental quality and the changing trends, accurately predict and forecast the occurrence probability of various types of environmental pollution accidents and improve capability on environmental analysis. Environmental data are an important means to improve the level of environmental information management in China. Big data can correlate seemingly unrelated and fragmented information for analysis in the context of environmental management so as to find trends, identify problems and grasp the laws, so data become the basis for decision making, which can help to enhance the accuracy and effectiveness of ecological governance and improve the quality of government management decisions.

Smart Technologies are Both "Kaleidoscopes" and "Microphones" for Ensuring Information Disclosure and Public Participation

Public participation is a strong guarantee of ecological governance. The development of smart technologies is conducive to the public access to environmental information in a timely manner, facilitating public participation in ecological governance decision making and supervision of ecological governance effects.

Firstly, smart technologies help to build a diversified platform and enhance the availability of environmental information. Environment and the public's vital interests are closely related. It is not the case that the public did not want to be involved in ecological governance in the past, but they could not actively participated because of limitations of environmental information technologies. The comprehensiveness and authenticity of environmental information disclosure is often difficult to guarantee. At present, enterprises under close state surveillance are connected online with environmental protection departments and these departments conduct supervisory surveillance quarterly and disclose the information to the public. Since 2015, air quality surveillance data have been released real time in 338 cities of China and all the data can be viewed in the website of Ministry of Environmental Protection. Some non-governmental environmental protection organizations also began to use mobile internet technologies to promote environmental information disclosure. For example, "Blue Map" developed by Beijing Public Environment Research Center can display not only information on the quality of air and rivers released by local authorities in China, but also the latest discharge data of enterprises via APP, website and other channels. Mobile internet software tools can also be used to check air quality data including PM 2.5 any time, which has become the habits of the public. The disclosure of environmental information allows the public to access environmental data anytime, anywhere and provides the necessary support for informed participation of the public.

Secondly, the creation of diversified channels have been created to stimulate enthusiasm of public participation. The rise of new media, especially microblog, WeChat and other popular social media, helps the public to exercise their rights of supervision more flexibly and conveniently and provides real time channels and broader platforms for the public to get involved in ecological governance. The concept of everyone being an observer, supervisor and environmental protector has become a reality. A variety of public interest micro projects in environmental protection have been carried out in new media. For example, Snapshots on Pollution in My Hometown, Snapshots on Black Chimneys and other activities have become hot topics in microblog, raising public awareness on supervision and stimulating public enthusiasm in participating in ecological governance. Government departments can also analyze and integrate the fragmented data uploaded by the public. In addition, the massive data produced in social media can help government departments to understand public needs so as to provide differentiated and refined public services and improve public awareness and experience in ecological governance.

第十四章　全球气候治理的经验与启示

冰川、冰盖融化，海平面上升，极端天气加剧，生物多样性减少，近海生态系统遭到毁灭性破坏，新发传染病增加……全球气候变化虽不足以决定历史走向，但可能带来的灾难性后果却倒逼人类直面气候治理问题，因而潜移默化地影响着历史进程。气候问题的复杂性与紧迫性要求世界各国调整国家政策与发展方向，携手设计合理有效的国际合作机制，以保卫人类共同的也是唯一的地球家园。近年来，国际社会在应对气候变化治理体系、规则、路径、工具等方面取得了一定成效。气候治理的实践和经验不仅有助于各国取得更多共识，形成更大的合力，亦对其他领域的全球治理有着重大意义，同时也为我国生态治理提供了可资借鉴的启示。

全球气候治理的进展

1988年，联合国环境规划署和世界气象组织成立了气候变化政府间会议（IPCC），气候变化的国际合作自此拉开了序幕。1992年《联合国气候框架公约》（简称《公约》）通过，这是世界上第一个"全面控制二氧化碳等温室气体排放以应对气候变化"的国际公约，为气候治理的国际合作奠定了法律基础。1997年，149个国家和地区通过的《京都议定书》规定，2008～2012年主要工业发达国家的温室气体排放量要在1990年的基础上平均减少5.2%，这是人类历史上首次以法规的形式限制温室气体排放。2007年，"巴厘路线图"进一步确认在《公约》和《京都议定书》下的"双轨"谈判，为气候治理的关键议题确立了明确议程。随后，2009年通过的《哥本哈根协议》确定了《京都议定书》一期承诺到期后的后续方案——2012～2020年的全球减排协议。2011年，各方在德班会议上决定启动"德班平台"，旨在于2015年前形成适用于《公约》所有缔约方的法律文件或法律成果，作为2020年后各方的贯彻《公约》，加强减排以应对气候变化的依据。

2015年的巴黎会议上,《公约》近200个缔约方一致同意达成新的全球协议,为2020年后全球应对气候变化的行动做出安排。2016年4月22日,在《巴黎协定》开放签署首日,共有175个国家签署了这一协定,并于11月4日正式生效。《巴黎协定》作为不足一年便迅速在全球得到批准并生效的多边协定,堪称前所未有。至此,1992年《公约》、1997年《京都议定书》以及2015年《巴黎协定》这三个人类历史上应对气候变化里程碑式的国际法律文本共同形成了2020年后全球气候治理的格局。尽管在责任分担和规则的制定上还存在很多矛盾,但经过多年摸索,各国在气候治理领域已经形成良好的合作态势,减少温室气体排放、应对气候变化已成为全球共识,包括发展中大国在内的主要国家都在积极地推进实质性减排。

全球气候治理的经验

气候治理最大的特点在于主体多元化及背后的利益多元化。因此,协调不同主体间的利益分配是推动全球气候治理的关键所在。以美国为首的某些发达国家虽背负气候变暖的历史责任,但囿于节能减排对经济发展的阻碍,相比因减排成本较低而积极主动的欧盟而言,历来抵触实质性减排。中东的石油大国出于维护石油产业的考虑,也反对实质性减排。而在海平面上升的威胁之下,欠发达国家中的岛国却一直强烈要求严格控制温室气体排放以确保国家安全。在这样的背景下,平衡多元利益诉求尤为重要。全球气候治理的经验主要体现在以下几个方面。

体系化规则建设有序引导气候治理。在国际层面,气候治理规则因《公约》而起,随《京都议定书》而兴,至《巴黎协定》而盛,它们的生效标志着气候治理规则的制度化和规范化。包括二十国集团(G20)、金砖五国、气候与清洁空气联盟(CCAC)在内的国际组织纷纷达成共识、加强合作,气候治理领域已逐步形成了以《公约》为主,其他小多边、双边、区域、地方等机制百花齐放的局面。在国家层面,在共同应对气候变化的进程中,越来越多的国家将应对行动和机制纳入自身法律和政策体系当中。2015年6月发布的《全球气候法规研究报告》显示，75个国家和欧盟已制定立法或政策框架来减缓气候变化。体系化的治理规则已成为各国共同实现利益最大化的有力保障。

国家自主贡献（Intended Nationally Determined Contributions，INDC）减排模式灵活推进气候治理。2013年的华沙大会要求各缔约方启动2020年后的自主贡献预案。《巴黎协定》则明确要求建立国家自主贡献机制，即各国可以根据自身国情、能力和发展阶段来提出各自的自主贡献目标，调整应对气候变化的措施和方法。与《京都协定书》"自上而下"的强制减排模式不同，国家自主贡献模式是"自下而上"地确定各国减排责任。除此之外，发达国家仍须继续带头减排，并加强对发展中国家的资金、技术和能力建设支持，帮助后者适应气候变化，从而避免了某些发达国家利用自主贡献而逃避减排责任的情形。如此安排不仅有助于鼓励所有国家采取相应行动，也有利于各缔约方订立切实可行的目标，有效履行其减排承诺。这样极具灵活性和包容性的做法充分考虑了各国的承受能力，促进新型动态减排分配机制和科学监督机制的形成，既尊重国家差异，又强化了所有国家的共同行动，充分反映出国际社会对于气候变化的统一认识与积极意志。

碳排放交易制度有效助力气候治理。多元利益冲突意味着不减排是最符合各国自身利益的选择。作为打破这种"囚徒困境"的有效政策工具，国别间的碳排放交易制度应运而生。1997年，随着《京都议定书》的订立，碳排放权正式成为国际商品并可通过市场进行流动与配置。该制度对大气进行了产权界定，通过控制市场上可排放的总量，为全球排放设定许可并依照一定标准在世界各国进行分配，确保"获益者担责"，以有效规制碳排放，实现全球减排目标。这一机制一方面为超量排放设立成本，鼓励节能减排，另一方面灵活调整各国之间的排放配额，让碳排放较多的发达国家承担更多的治理经费，是市场机制有效推动全球减排的典型体现。

气候治理实践的启示

全球气候治理的经验或可为推进我国生态文明建设领域国家治理体系和

治理能力现代化，思考和探索未来全球治理模式提供宝贵启示。

不断提升生态治理领域法治水平。应当从立法层面尽快推动相关法律、法规、条例的制定与完善，健全跨区域、跨流域治理规则体系。借鉴全球气候治理的经验，强调不同地区互相协助、共同面对的理念。在治理规则的制定过程中，应当落实听证环节，完善信息公开与通报制度，让企业、非政府组织和公民等多类主体实现更具广度和深度的参与，以充分听取各方诉求，真正考量各方利益。充分发挥地方智慧，鼓励地方积极探索区域、流域合作的多样化机制。

创新地区生态责任分配机制。在全球气候治理问题上，创新的减排承诺机制大大提高了各国的参与度，并在一定程度上提高了治理方案的可行性。究其原因，本质上是由于该分配机制充分考虑并尊重了各国之间的客观差异。我国在解决跨地域、跨流域等生态环境问题时，应当综合考量各地区之间的具体状况、发展差异以及社会文化等因素，建立相应的对话机制，充分听取地方意见，鼓励地方根据自身发展状况提出相应提案。例如在流域生态补偿实践中，上游地区的用水情况以及污染将对下游地区产生极大的影响，如要照顾下游的供水质量，上游地区的发展空间必然在一定程度上受到限制，因此建立上下游平等、稳定、有效的对话机制，合理平衡上下游地区之间的责任与利益分配变得至关重要。

充分发挥市场机制在生态治理中的作用。以往许多环境污染和资源耗竭都是伴随着粗放型生产方式而发生，粗放型生产方式意味着资源的过度消耗和污染物的过度排放。过度消耗和排放的症结在于生态外部性为大众共担，经济利益却被少数人享有。欲遏制这种消极态势，必须重视市场机制的作用，将生态外部性内部化。实行资源有偿使用制度和生态补偿制度，使排放权成为有价商品，通过市场机制优化排放权的配置。在此基础上，环境资源就有了可衡量的价值标准，收益和责任便联系在了一起，市场主体就会有主动治理的动力。

Chapter 14
Experiences and Inspiration from Global Climate Governance

Melting glaciers and ice sheets, rising sea levels, extreme weather, decreasing biodiversity, devastating damages to coastal ecosystems and increasing new infectious diseases, etc. are phenomena of global climate change which have not aroused enough attention to determine historical direction, but the disastrous consequences climate change has possibly brought about have forced the mankind to take it seriously, thereby unconsciously influencing historical process. The complexity and urgency of climate change requires that all countries in the world adjust their national policies and development directions and work together to design rational and effective international cooperation mechanisms to defend the common and the only homeland of human beings. In recent years, the international community has achieved certain results in developing governance systems, rules, pathways and tools in response to climate change. The practice and experience of climate governance can help countries to achieve more consensus and greater synergy and are significant to global governance in other fields. Meanwhile, they can provide useful reference to China's ecological governance.

Progress of Global Climate Governance

In 1988, the United Nations Environment Program and the World Meteorological Organization set up Intergovernmental Panel on Climate Change (IPCC) and international cooperation on climate change has since kicked off. In 1992, *United Nations Framework Convention on Climate Change* (UNFCCC) was adopted and this is the first international convention on "global control of greenhouse gas emissions such as carbon dioxide to cope with climate change" in the world, laying the legal foundation for international cooperation in climate governance. In 1997, *Kyoto Protocol* passed by 149 countries and regions stipulates that during 2008 ～ 2012, major industrialized countries should achieve a 5.2% reduction of greenhouse gas emissions based on the emission level in 1990. For the first time in human history, greenhouse gas emission was restricted in the form of a regulation. In 2007, *Bali Roadmap* further confirms the dual-track negotiations under UNFCCC and *Kyoto Protocol*, setting a clear agenda for key issues of climate governance. Subsequently, the *Copenhagen Accord* adopted in 2009 clarified the global emission reduction agreement for 2012 ～ 2020 after Phase I commitment made in *Kyoto Protocol* expired. In 2011, the parties decided at the Durban Conference to launch the *Durban platform* with a view to developing legal documents or achieving legal outcomes applicable to all Parties to the Convention by 2015 which can serve as a basis for all parties to implement the *Convention* by strengthening emission reduction in response to climate change.

At the Paris Conference in 2015, nearly 200 Parties to the Convention agreed to a new global agreement to make arrangements for global response to climate change beyond 2020. On April 22, 2016, 175 countries signed *the Paris Agreement*, which entered into force on 4 November. As a multilateral agreement, *the Paris Agreement* was ratified and entered into force on a global basis for less than a year, which is unprecedented. At this point, the three international legal documents in response to climate change in human history, namely, the 1992 *Convention*, the 1997 *Kyoto Protocol* and the 2015 *Paris Agreement*, have formed a pattern of global climate governance beyond 2020. Although there are many contradictions in sharing responsibilities and formulating rules, after years of exploration, countries have conducted remarkable cooperation in the field of climate governance and global consensus has been reached on reducing greenhouse gas emissions and responding to climate change. Major countries including developing countries are actively promoting substantial emission reductions.

Experience from Global Climate Governance

The greatest feature of climate governance is the diversification of the stakeholders and interests behind. Therefore, the coordination of the distribution of benefits among different stakeholders is the key to promoting global climate governance. Some developed countries, led by the United States, have the historical responsibility of climate warming, but they, unlike EU who are active in emission reduction due to relatively low cost, are reluctant to carry out substantial emission reduction because energy conservation and emission reduction could hamper economic development. The major oil-producing countries in the Middle East are also opposed to substantial emission reduction because they want to protect the oil industry. Some underdeveloped island countries, which are under the threat of sea level rise, have been strongly demanding strict control of greenhouse gas emissions to ensure national security. In this context, it is particularly important to balance the needs of multiple interests. The experience of global climate governance is mainly reflected in the following aspects.

Development of systematic rules can guide climate governance. At the international level, the rules for climate governance started with the *Convention*, improved with *Kyoto Protocol* and matured with *Paris Agreement*. Their entry into force marked the institutionalization and standardization of climate governance rules. The international organizations including the G20, the BRICS and the Climate and Clean Air Coalition (CCAC) have reached consensus and strengthened cooperation. In the field of climate governance, there are many multi-lateral, bilateral, regional and local mechanisms in addition to the guiding Convention. At the national level, in the process of coping with climate change, more and more countries have incorporated response actions and mechanisms into their own legal and policy systems. *The Report on Global Climate Regulations Research* released in June 2015 showed that 75 countries and the European Union have enacted legislation or policy frameworks to mitigate climate change. Systematized governance rules have become a powerful guarantee for all countries to jointly maximize profits.

The emission reduction model of Intended Nationally Determined Contributions (INDC) flexibly promotes climate governance. The Warsaw Conference in 2013 required all parties to initiate self-contribution plans after 2020.

The Paris Agreement explicitly requires the establishment of a national mechanism for independent contributions, that is, countries can propose their own independent contribution goals according to their own national conditions, capabilities and stages of development and adjust the measures and methods for dealing with climate change. Unlike the "top down" model of compulsory emission reduction under *Kyoto Protocol*, the national self-contribution model is a "bottom-up" determination of emission reduction responsibilities in various countries. In addition, developed countries still need to continue to take the lead in reducing emissions and strengthen support for developing countries regarding funds, technologies and capacity to help them adapt to climate change, thus preventing some developed countries from using their own contributions to evade emission reductions. Such an arrangement not only helps encourage all countries to take corresponding actions, but also helps parties to set practical goals and effectively fulfill their emission reduction commitments. Such a highly flexible and inclusive approach fully takes into consideration the affordability of countries and promotes the formation of new dynamic emission reduction allocation mechanisms and scientific monitoring mechanisms. It not only respects national differences but also strengthens joint actions of all countries, which fully reflects the unified understanding and positive will of international community for climate change.

The carbon emission trading system effectively facilitates climate governance. Multiple conflicts of interest mean that zero emission reduction is in the interests of each country. As an effective policy tool to break this prisoner's dilemma, a carbon emission trading system between countries has emerged. In 1997, with the signing of *Kyoto Protocol*, carbon emission rights officially became international commodities and could be circulated through the market. The system defines the property rights of the atmosphere, controls the total amount of emissions in the market, sets a license for global emissions and allocates them to countries in the world according to certain standards, ensuring that "beneficiaries are responsible" so as to effectively regulate carbon emissions and achieve global emission reduction targets. This mechanism, on the one hand, establishes costs for over-emissions and encourages energy-saving and emission-reduction. On the other hand, it flexibly adjusts the emission quotas among countries and allows more carbon-intensive developed countries to shoulder more governance funds. It is typical embodiment that a market mechanism can effectively promote global emission reduction.

Inspirations from Climate Governance Practices

The experience of global climate governance may provide valuable inspirations for advancing the modernization of the state governance system and governance capabilities in the field of ecological civilization construction in China and for exploring future global governance models.

Continuously improve the rule of law in the field of ecological governance. It is necessary to promote the formulation and improvement of relevant laws, regulations and regulations as soon as possible from the legislative perspective and improve the cross-region and inter-basin governance rules. Drawing on the experience of global climate governance, we should emphasize the concept of mutual assistance and cooperation in different regions. In the process of formulating governance rules, we should implement the hearing system and improve the information disclosure and notification system so that enterprises, non-governmental organizations, the public and other types of players can achieve greater participation to fully listen to the demands of all parties and consider interests of all parties. Local wisdom should be given to full play and local governments should be encouraged to actively explore the diversified mechanisms of regional and river basin cooperation.

Innovate the regional ecological responsibility distribution mechanism. Regarding global climate governance, innovative emission reduction commitment mechanisms have greatly increased the participation of various countries and have, to a certain extent, improved the feasibility of governance programs. The reason is that the allocation mechanism fully considers and respects the objective differences among countries. In solving the problems of trans-regional and inter-basin eco-environmental problems, China should comprehensively consider the specific conditions, development differences and social and cultural factors among various regions, establish a corresponding dialogue mechanism, fully listen to local opinions and encourage local development according to their own conditions. For example, in the practice of watershed ecological compensation, water use and pollution in the upper reaches will have a great impact on the downstream areas. If the downstream water supply quality is to be taken care of, development in the upper reaches must be limited to a certain extent. An equal, stable and effective dialogue mechanism that balances responsibilities and benefits between upstream and downstream regions becomes crucial.

Give full play to the role of market mechanism in ecological governance. In the past, much environmental pollution and resource depletion have been accompanied by extensive production methods, which means excessive consumption of resources and excessive discharge of pollutants. The crux of excessive consumption and emission is that ecological externality is shared by the general public and economic gains are enjoyed by a small number of people. To curb this negative trend, we must attach importance to the role of market mechanisms and internalize ecological externalities. The system of paid use of resources and the system of ecological compensation will be implemented so that the emission rights will become a valuable commodity and the allocation of emission rights will be optimized through market mechanism. On this basis, there will be measurable value standards for environmental resources and benefits and responsibilities will be linked together. Market players will have the motivation to actively take measures.

第十五章　谱写绿色丝绸之路新篇章

"一带一路"是我国应对国际新形势、统筹国内国际两个大局、探索全球治理新模式的重大战略决策。"一带一路"横跨亚非欧三大洲，涉及沿线60多个国家和地区、44亿人口及21万亿美元的经济总量，范围之广、规模之大、影响之深史所罕见。在推进"一带一路"的过程中，必须将生态环保和绿色发展作为重要支撑。绿色丝绸之路本质上是将生态文明理念融入"一带一路"中，夯实"一带一路"的生态环境基础，倡导沿线各国走绿色发展之路。这对于改善沿线国家的生态环境具有积极意义，也是我国充分发挥绿色领导力、参与全球治理的重要举措。

建设绿色丝绸之路面临的挑战

1."一带一路"沿线地理环境复杂，生态脆弱。"一带一路"跨越高原、森林、荒漠、草原、农田、海洋等多类生态系统，区域差异大，环境本底差。哈萨克斯坦、沙特阿拉伯等中亚、西亚地区，干旱少雨，土地沙漠化形势严峻；俄罗斯等高纬地区，植被生长缓慢，生态恢复能力差；印度、新加坡等南亚和东南亚地区，雨林锐减，城市化压力大，面临生物多样性不断减少的威胁；印度尼西亚等部分海域地区，海洋生态保护压力较大。

2."一带一路"沿线发展水平较低，人口密度高，经济发展与生态保护之间矛盾突出。"一带一路"沿线区域面积不到世界的40%，但人口却占世界的近64%。沿线国家绝大多数仍是发展中国家，生产和消费需求大，经济发展方式粗放，对水、石油、矿产等资源的开采力度大，恶化了原本较为敏感脆弱的生态环境，增大了区域生态安全威胁。一些国家甚至出现了经济进则生态退，欲生态进须经济退的两难局面。

3."一带一路"带来潜在的环境压力。随着"一带一路"的推进，沿线国家面临的生态风险和环境压力将会更加突出。目前，在"一带一路"项目建设和资源开发过程中，存在大气污染、水土流失、尾矿污染等生态风险。同时，清洁能源产业合作和基建工程也可能带来一定的环境影响：太阳能发电站的维护，需要用足够的水清洗电池板，耗水量巨大；水电大坝的建设则可能对水道生物多样性造成威胁。此外，产业项目涉及区域较广，可能穿越或隔离生境，造成物质、能量、信息的交叉、阻挡或限制。如交通设施的建设可能改变土地利用格局，破坏覆盖植被，阻碍动物迁徙。

4."一带一路"沿线国家间存在资源权属冲突和环保标准不统一。一方面，沿线各国资源权属争议不休。西亚、中亚、南亚、东南亚等部分地区资源短缺，长期处于政局动荡、主权争议、资源争夺的环境下，尤其是在石油、天然气、稀有金属、水分配等领域的分歧与矛盾始终未能解决。另一方面，沿线各国环保标准不一。如俄罗斯、哈萨克斯坦的水质标准高于中国，东盟多国水中重金属控制标准严于中国。而中国在火电、钢铁等行业的排放标准高于国际标准。"一带一路"绿色发展应当实行国际标准、国内标准还是当地标准尚无定论。

明确绿色丝绸之路主要布局

1.科学规划绿色互联互通体系。在交通体系建设方面，应当将环境影响纳入港口、场站布局的考虑，积极发展海铁联运、内河运输等能耗较少、环境友好的运输方式，共同打造通畅安全、绿色高效的运输大通道。在能源基础设施建设方面，应当加大各国在跨国油气管线、发电站建设和跨境电网互联的合作力度，注重推进光伏、风能等绿色能源基建合作。在通信网络建设方面，应优化核心网供电制冷系统和基站供配电系统的能耗，积极开发利用新能源，对通信网络进行全方位绿色变革，打造绿色丝绸之路的信息桥梁。

2.推进绿色产业及技术合作。在农业方面，建立绿色农产品生产加工基地，发展绿色农产品电子商务，引进优良种质资源与人才技术，形成绿色农产品深加工产业链。在工业方面，要进一步加强绿色工业园区作为我国和其他沿线国家开展绿色工业合作的主要模式的地位。以绿色基础设施建设为重点，开展对可再生水、地源热泵等绿色技术的利用，尽量降低生产的物耗和能耗。在服务业方面，挖掘沿线民族特色浓厚、丰富多样的旅游资源，组建沿线旅游机构联盟，积极培育生态旅游。

3. 构建绿色金融体制。通过股票、债券、贷款、私募投资、保险等金融工具严格控制污染性投资，并配套政策性出口信用保险和银行之间的银保合作机制，将社会资金引导到节能环保、清洁能源等绿色产业，从而实现环境可持续发展的投融资模式，助力建设绿色丝绸之路。针对产业项目的绿色风险建立一套完整的评价体系和技术指标，建立第三方绿色项目评估市场。加强国际合作，建立区域绿色基金，强化亚投行等专业国际组织对绿色金融的支持，使"赤道原则"被更多金融机构采纳，充分审慎衡量融资项目中的环境风险。

4. 建设跨界生态廊道和保护区。通过建立跨界生态廊道和保护区，为自然生态和生境的保护确立更广阔的邻接面积、更深入的联合防控，实现对"一带一路"的全面生态保护。通过双边、多边的磋商会谈，在高层行动上，就跨界生态廊道和保护区的建设达成战略合作意向，就联合主体、监督实体、紧急情况下的互助等宏观内容签订框架协议；在地方行动上，相邻国的保护区签订生态保护相关的地区合作文件，在规划、监测、管理、研究、保护等方面达成更为具体的合作内容，或通过非政府组织对跨界行动提出倡议并给予技术、贷赠款等方面的协调和支持。

完善绿色丝绸之路治理体系

1. 增进绿色发展的理念共识，加强沿线国家和地区之间的交流合作。绿色丝绸之路能够极大地推动绿色发展和生态文明理念的传播，有利于构建生态环境责任共同体，最大限度地减少生态环境影响，让各国人民享受更多绿色"红利"。加强国际交流合作有助于推广绿色发展理念，凝聚共建绿色丝绸之路的合力，要积极完善双边、多边交流合作机制，重点加强绿色贸易、绿色供应链、环保产品与服务业合作。充分发挥"丝路国际论坛""中阿环境合作论坛"等对话平台的作用。

2. 深化绿色产能合作，完善绿色基础设施建设。妥善处理产能合作和基建过程中的生态环境风险，制定事前生态保护规划指南，为产能和基建的合作提供绿色指导。光伏、钢铁、玻璃、水泥等领域的产能合作必须充分了解沿线国家和地区的环保标准及要求，将环境风险控制在生态环境容量之内。基础设施建设则需要加强环保规划、施工监理的衔接与配合，事前预防优先于事后治理。推进节能环保产业进行核心技术创新，实现绿色产业升级。深化在绿色交通、绿色能源、绿色金融、绿色信息等领域合作，推动绿色产业高质量输出。

3. 构建"一带一路"生态信息平台，统一的绿色发展要求和标准。生态信息平台的构建有助于消除生态信息不全面、不畅通、不对称的现象，加强各国之间环境保护信息的交流和生态环境基础数据的研究与分析，为解决绿色发展要求和标准的冲突和不统一提供信息支撑。统一的生态保护标准有助于衡量各国的生态治理效果，更好地推进跨区域生态治理。在国际标准化组织制定的ISO14000环境管理体系的基础上，结合各国国情，协商制定适用于"一带一路"沿线国家和地区的涵盖大气、水、土壤、生物、海洋等各方面的生态保护标准。

4. 完善资源环境绩效评估，鼓励多元主体深入参与。引入第三方专业机构对"一带一路"沿线国家资源环境绩效进行评估，为开展绿色发展领域的国际合作提供更明确的指导和依据。科学设定评估指标体系，增强数据的可靠性，确保评价结果的权威性和可比性。"一带一路"沿线多种族、多民族、多宗教、多文化的现实情况要求必须重视企业、机构、组织、个人等多方力量，增进沿线各国对绿色发展理念的认同，鼓励多元主体共同参与绿色丝绸之路的建设。

Chapter 15
A New Chapter for Green Silk Road

The Belt and Road Initiative (BRI) is a major strategic decision in response to new international situation with the purpose of exploring new patterns of global governance and planning as a whole domestic and international development. Stretching across the three continents of Asia, Africa and Europe, the Initiative covers over 60 countries and regions with a population of 4.4 billion and an economic aggregate of 21 trillion US dollars. The scope, scale and impact of the Initiative have been unprecedented in history. In the process of implementing the Initiative, ecological protection and green development will be the guiding principles. In essence, the concept of ecological civilization will be integrated into the green silk road, laying a solid ecological and environmental foundation for the Initiative. Countries along the belt will also take on the road of green development. The initiative is of positive significance in improving ecological environment in these countries. It is also an important measure China takes to fully display her green leadership and engage in global governance.

Challenges Facing the Development of Green Silk Road

1. The geographical environment is very complex with fragile ecology along the belt and road. The Initiative covers many different types of ecological systems including plateau, forest, grassland, farmland and ocean, etc. with vast regional differences and poor environmental conditions. Countries in Central Asia and West Asia such as Kazakhstan, Saudi Arabia, etc. are dry with little rainfall and severe land desertification. Russia and other countries in high-latitude areas have slow vegetation growth and poor ecological restoration. India, Singapore and other countries in South Asia and Southeast Asia are faced with deceasing rain forest, high pressure of urbanization and serious threat of reducing bio-diversity. Indonesia

and other countries in coastal areas are under great pressure of protecting marine ecology.

2. Countries along the belt and road have low economic development level and high population density. Prominent contradictions exist between economic development and ecological protection. The area of the region accounts for less than 40% of the world total, yet its population accounts for near 64% of the world. The majority of the countries are still developing countries with tremendous demand for production and consumption and extensive economic development models. There countries excessively exploit their resources such as water, oil and mining, etc., which has worsened the sensitive and fragile ecological environment and increased threats to regional ecological security. Some countries are caught in the dilemma between economic development and ecological deterioration.

3. The Initiative can cause potential environmental stress. With the launching of the Initiative, relevant countries will face more prominent ecological risks and environmental pressure. At present, there exist ecological risks such as air pollution, soil erosion and tailing pollution in the process of project implementation and resource development, meanwhile, cooperation in clean energy industry and infrastructure projects can also generate certain environmental impact: The maintenance of solar power stations needs sufficient water to clean the battery panels. The construction of dams for hydropower stations may threaten biodiversity in waterways. In addition, industrial projects crossing vast regions may traverse or isolate habitats, resulting in overlapping, obstructions or restrictions of matter, energy and information. For example, constructions of transportation facilities may change the pattern of land use, destroy vegetation coverage and impede animal migration.

4. There are conflicts of resource ownership and differences in environmental standard among the belt and road countries. On the one hand, these countries constantly dispute over resource ownership. West Asia, Central Asia, South Asia and Southeast Asia and other regions are short of resources. Some countries in these regions have long suffered from political turmoil, sovereignty disputes and resource scrambles. In particular, their differences and conflicts over allocation of

oil, natural gas, rare metals and water, etc. have not been resolved. On the other hand, environmental standards differ in the belt and road countries. For example, water quality standards in Russia and Kazakhstan are higher than that in China. The heavy metal control standards in many of ASIASN countries are stricter than those in China. China's emission standard in thermal power, iron & steel and other industries is higher than international standard. Whether international standards, domestic standards or local standards should be adopted for green development remains undecided.

Formulating Blueprints for Green Silk Road

1. Plan green and interconnected systems in a scientific way. Regarding transportation systems, environmental impact should be included in planning the layout of ports, airports and stations. Environmental-friendly transportation modes with less energy consumption like sea-rail combined transportation and inland water transportation, etc. should be actively developed to build safe, smooth, green and efficient transportation arteries. Regarding energy infrastructure, cooperation should be enhanced among countries in building cross-nation oil and gas pipelines, constructing power stations and connecting cross-region power grids and efforts should be made to promote cooperation in PV, wind power and green energy infrastructures. Regarding communication network, power supply and refrigeration system of core networks and energy consumption of power supply and distribution system in base stations should be optimized. New energy should be actively developed to carry out comprehensive green transformation of communication networks so as to build information bridges for the green silk road.

2. Promote green industry and technical cooperation. In the field of agriculture, we should establish green agricultural production and processing bases, develop E-commerce for green agricultural products, introduce fine germplasm resources, talent and technology and build deep processing industrial chains for green products. In terms of industry, it is necessary to build green industrial parks which can be used as the main model for China and other BRI countries to carry out green

industrial cooperation. We should focus on green infrastructure cooperation, utilize such green technologies as renewable water and ground source heat pumps and minimize material consumption and energy consumption in the production process. Regarding service industry, we should explore rich and diverse tourism resources with strong national features by establishing tourism alliances and actively fostering the development of eco-tourism.

3. Build a green financial system. We should strictly control investment in polluting industries through stock, bonds, loans, private equity, insurance and other financial instruments and guide private capital to green industries such as energy-conserving, environmental-friendly and clean energies through the bank-insurance cooperation mechanism between policy-oriented export credit insurance companies and banks in order to develop environmentally sustainable financing modes and support the construction of a green silk road. A complete set of evaluation systems and technical indicators and a third-party green project evaluation market should be established to control the green risks of industrial projects. International cooperation should be enhanced to set up regional green funds and the role of professional international organizations such as Asia Infrastructure Investment Bank (AIIB) should be strengthened to support green finance. Equator Principles should be adopted by more financial institutions to fully measure environmental risks in financing projects.

4. Build cross-border ecological corridors and protected areas. Through the establishment of cross-border ecological corridors and protected areas, a broader adjoining area and more in-depth joint prevention and control for the protection of natural ecology and habitats will be established to achieve comprehensive ecological protection for the BRI regions. Strategic cooperation intentions on the establishment of cross-border ecological corridors and protected areas can be reached through bilateral and multilateral high-level consultations and negotiations. Framework agreements can be signed on general items such as joint operations, supervisory entities and mutual assistance in emergencies, etc. On local levels, protected areas in neighboring countries can sign regional cooperation documents related to ecological protection including specific contents such as planning, monitoring, management,

research, protection, etc. NGOs can propose initiatives for cross-border operations and governments of different countries can provide coordination and support in the form of technical assistance, loans and grants.

Improving the Governance System for the Green Silk Road

1. Promote the consensus on the concept of green development and strengthen exchanges and cooperation between BRI countries and regions. The green Silk Road can greatly promote the concept of green development and ecological civilization. It is conducive to building a community with ecological responsibility, minimizing the impact of the ecological impacts and allowing people of all countries to enjoy more green "dividends". Strengthening international exchanges and cooperation will help promote the concept of green development and enhance synergies in building a green Silk Road. We must actively improve the bilateral and multilateral exchange and cooperation mechanisms, focusing on strengthening green trade, green supply chains, environmental-friendly products and service industry cooperation. We should give full play to the role of dialogue platforms such as Silk Road International Forum and China-Arab Environmental Cooperation Forum.

2. Deepen green capacity cooperation and improve green infrastructure construction. We will properly handle the ecological and environmental risks in the process of capacity cooperation and infrastructure construction and formulate ecological protection planning guidelines in advance to provide green guidance for cooperation in production capacity and infrastructure construction. We must fully understand the environmental protection standards and requirements of BRI countries and regions when carrying out capacity cooperation in photovoltaic, steel, glass, cement and other fields and control environmental risks within the ecological environment capacity. Regarding infrastructure construction we must strengthen environmental protection planning and construction supervision. Prevention is better than cure. We must promote core technological innovations in energy-saving and environmental protection industries so as to upgrade green industries. We must deepen cooperation in the fields of green transportation, green energy, green finance,

and green information to promote the high-quality output of the green industries.

3. Build an ecological information platform and harmonize requirements and standards on green development. The construction of an ecological information platform helps to eliminate the phenomenon of incomplete, impeded and asymmetrical ecological information, strengthen the exchange of environmental protection information and the research and analysis of basic ecological data in different countries and provide information support to resolve the conflict caused by different requirements and standards on green development. A unified ecological protection standard helps to measure the ecological governance of countries and better promote cross-regional ecological governance. Based on the ISO14000 environmental management system formulated by the International Organization for Standardization and in accordance with the national conditions of various countries, ecological protection standards on air, water, soil, biology, and oceans, etc. should be formulated applicable to BRI countries and regions through negotiation.

4. Improve resource and environmental performance assessment and encourage multiple entities to fully participate. The introduction of third-party professional organizations to assess the national resources and environmental performance provides more clear guidance and basis for the launch of international cooperation in green development. We should scientifically design an evaluation index system to enhance the reliability of data and ensure the authoritativeness and comparability of evaluation results. The BRI countries have different ethnic groups, nations, religions and cultures, which requires that enterprises, institutions, organizations, individuals, etc. should reach consensus on the concept of green development. Multiple entities should be encouraged to participate in green Silk Road construction.

第十六章　国家公园：奏响生态治理现代化新乐章

十八届三中全会从国家治理体系和治理能力现代化的改革总目标出发，提出"坚定不移实施主体功能区制度，建立国土空间开发保护制度，严格按照主体功能区定位推动发展，建立国家公园体制"。在推进生态文明建设、实现生态领域治理体系和治理能力现代化的进程中，建立国家公园体制是设定资源消耗上线，划定生态保护红线，严守环境质量底线，构建国土生态安全空间的重大举措，是整合现有保护地类型，形成科学统一、与国际接轨的保护地体系的强大动力。通过建立国家公园对本国具有代表性的生态系统、自然资源及景观、文化遗产等予以重点、有效保护，已成为国际社会的趋势。

国家公园：正本与清源

我国"国家公园"类型多样，自然保护区、风景名胜区、世界文化自然遗产、国家森林公园、国家湿地公园、城市湿地公园、国家沙漠公园、国家矿山公园、国家地质公园、国家重点公园及水利风景区等十几种不同名称的"公园""保护区"重叠司空见惯，多块牌子的背后呈现出管理体系的混乱和部门利益的博弈。在实践中不免困惑于归谁管，怎么管。因此，建立国家公园体制，需要先正本清源。

事实上，"国家公园"（national park）由来已久，已有140多年的历史。1872年，世界上第一个国家公园——美国黄石国家公园正式创立。虽然各国国家公园定义并不统一，模式不尽相同，但国家公园的百年实践也逐渐形成了共识：国家公园不同于以游憩为主要导向的一般意义上的公园，而是以保护为优先考虑，兼顾公众享有。国家公园具有明确的限定要素，并不能仅从名称判断。结合IUCN《保护地管理应用指南》，域外成熟的制度运行和实践经验，以及我国现实国情，我国的"国家公园"应当包括四个要素：第一，资源要素。国家公园内的生态系统、自然和文化景观应当具有国家代表性，且面积要足够大、生态质量要足够高，以使得其内的生态进程得以持续进行。第二，功能要素。国家公园应当有一定区域为公众提供娱乐、游憩、学习、教育的机会。第三，国家要素。中央在国家公园的管理、财政等方面应当承担重要甚至主导作用。第四，程序要素。成为名正言顺的国家公园需要符合相应的程序要件，即通过设立程序，由管理部门根据相应的标准进行指定，并实施规划管理。

2006年以来，云南、黑龙江、浙江等地都进行过"国家公园"地方层面的试点。各地国家公园的探索实践，对于整合政出多门的保护地管理现状，推行有效的保护地体系，保护有重大生态价值的自然生态系统具有先行意义。然而，由于发展旅游能够带来巨大的经济效益，并且社会公众对于保护地所能带来的教育、游憩的需求强烈，在此背景下，地方对于设立国家公园试点的积极响应可能更看重国家公园所能带来的品牌效应和经济效益。以探索国家公园试点为名，行资源开发利用创收之实，对环境、生态的破坏不可避免，很难体现出国家公园的真正特性。

为了纠正乱象丛生的试点实践，有序建立国家公园体制，2015年以来，中央进行了统一部署：13个部委联合出台试点方案，由国家发改委牵头成立工作协调机制，遴选9个地区为试点。2015年，青海三江源、浙江开化及湖北神农架的国家公园试点正式启动。作为我国第一个国家公园体制试点，三江源国家公园的建设着力突破原有体制的藩篱，解决"九龙治水"和监管执法碎片化问题，构建归属清晰、权责明确、监管有效的生态保护管理体制机制，为我国其他国家公园试点建设树立了标杆与基准。

生态治理：同音且共律

推进生态治理现代化、加强生态文明建设是缓解当前严峻的生态环境危机的必由之路，更是实现国家、民族永续发展的战略选择。建立国家公园体制，将在理念、体制及行动三方面带动生态治理现代化，助力生态文明建设。

理念上，国家公园体制的成功建立将会极大地提升和增强国家认同感和民族自豪感，凝聚全民生态保护的共识，汇集生态文明建设的合力。缺少保护生态的共同信念，生态治理难以得到全社会多元主体的合力推进，更毋言其现代化。建立国家公园体制的实践则为这种共识的形成提供了一剂良方。国家公园不仅是一种重要的保护地类型，更代表一国最具代表性的自然美景和文化遗迹，通过让人们全身心地接触美丽的自然景观和瑰丽的文化遗迹，生态保护意识落地生根，环境保护理念日益增强，从而产生对自然的爱护之情和对国家文化的赤子之心，并将逐渐体会到保护国家公园功在当代、利在千秋。这种直接通过自然和文化之美所汇集的生态保护意识能够成为推动生态治理现代化的强大精神力量。

体制上，建立以国家公园为统领的科学、规范、统一、高效的保护地体系，能够成为推动生态治理现代化的先导力量。一方面，完善保护地体系属于生态治理的典型内容。建立国家公园体制并非仅指设立国家公园这一类保护地，而是在对我国碎片化、高度重叠的各类保护地进行整合的基础上，充分考虑自然的整体性，用国家公园体制予以统领，解决人为碎片化生态系统的问题。另一方面，国家公园体制的建立又能够为解决生态治理现代化的关键问题提供深入研究的基础。解决国家公园内自然资源权属问题，能为深化自然资源资产产权改革，统一管理山水林田湖提供经验。明晰国家公园内保护及利用的界限，能够为科学设定资源消耗上线，划定生态保护红线，严守环境质量底线，构建国土生态安全空间提供解决思路。以国家公园为统领的类型多样全面，设立科学合理、责权利清晰、监管有力、资金投入稳定、多元主体协同治理的保护地体系，将成为率先实现生态治理现代化的突破口。

行动上，国家公园建设中多元主体参与特征显著，是生态领域由"管理"向"治理"转变的典型。多国成熟的国家公园治理实践表明，国家公园从规划、建立到维护的整个动态过程，都需要国家、企业及公众多方力量的协同行动。中央需要做好对国家公园体制的顶层设计，承担主要及兜底的运营经费。地方则要落实配套制度，予以管理、维护等具体实践层面的支持。企业和其他社会组织，可以通过特许经营、社会捐赠、协助培训等为国家公园的发展提供资金和智力支持。而公众，尤其是周边居民，不仅有机会成为国家公园的工作人员或志愿者，而且可以通过参观国家公园、支付相应的门票费用，做出自己的经济贡献。多元主体积极参与生态治理，在一定条件下，是国家生态治理现代化能够实现的重要保证。

法治指挥：奏出新乐章

长期以来，整合保护地体系难以推进的主要原因在于保护地法律依据的部门化和碎片化。虽然迄今已出台《中华人民共和国自然保护区条例》《风景名胜区管理条例》《国家级森林公园管理办法》《国家重点公园管理办法》等多个法规、部门规章及规范性文件，但是其适用范围受限于单个保护地类型，法律效力较低，实践中各行其是，缺乏统筹协调功能，难以实现建立国家公园体制的整体构想。因此，应通过分析和总结国家公园体制试点中的实施困境和先行经验，就国家公园立法的核心内容达成共识，制定专门的《国家公园法》，将国家公园体制的建立纳入法制化的轨道。

《国家公园法》总体思路应当包括五方面：第一，《国家公园法》是为了什么？这就需要在立法目的中正面回答保护优先于利用，避免出现含糊不清而使得保护优先的立法宗旨难以凸显。第二，国家公园是什么？法律定义需要在适当借鉴世界自然保护联盟（International Union for Conservation of Nature，IUCN）概念的基础上，全面考虑我国自然和人文资源的分布情况，突出面积、功能、国家及程序要素。第三，《国家公园法》怎么做？就基本原则而言，需要做到保护第一、保障公益、国家主导、科学指导。第四，《国家公园法》怎么用？本法的适用范围应当包括国家公园的设立、规划、保护、管理、利用、维护、监督等活动及行为。第五，谁对国家公园负责？这涉及不同主体的权利与义务：有关机构的职责应贯串整个过程，既涉及管理体制的安排，又涉及经费、监督等方面；原住民、公众等应当拥有对国家公园进入、游憩、享受的权利，也应承担不因其利用而打扰到国家公园维持自然状态的保护义务。

此外,《国家公园法》应当对国家公园的设立标准、资源权属、规划报告、管理体制、资金投入、经营方式、监督措施、监测评估机制等重要内容的核心要义予以明确的规定,对于这些主要制度再通过详细的法规、实施细则或者管理类、技术类的规范予以落实。充分考虑国家公园的生态整体性和国家代表性,明确国家公园的资源权属,确保国家公园得到稳定的资金投入,实现对国家公园集中、高效、统一的保护管理,使国家公园体制在全面、系统、详细的一整套法律规范体系的前提下得以建立。

以法治为指挥棒,在国家、企业、公众等多方主体的共同演奏下,才能吹响建立国家公园体制的号角,奏出生态治理现代化的和谐乐章。

Chapter 16
National Parks Play an Important Role in Ecological Governance

The Third Plenary Session of the 18th CPC Central Committee stated that "we should unswervingly implement the functional zoning system, build a territory spatial development and protection system, establish a national park system and promote development in strict accordance with positioning of functional zones". In the process of promoting ecological progress and modernizing ecological governance system and governance capabilities, establishing a national park system is a major measure to set the upper limits for resource consumption, draw a red line for ecological protection, strictly adheres to environmental quality bottom line and ensure national ecological security. It is also a powerful driving force in integrating existing protected areas and building a scientific and unified system of protective areas in line with international practices.

National Park: What is it?

China has many different types of "national parks" including natural reserves, scenic spots, world cultural heritage sites, national forest parks, national wetland parks, urban wetland parks, national desert parks, national mine parks, national geological parks, national key parks, water conservancy scenic areas and other parks with a dozen different names. The similar and overlapping names reflect the confusing management systems and the conflicting interests between different departments. It is really bewildering who are in charges of national parks and how they are managed. Therefore, we should clarify the definition and get to the bottom of the issue.

In fact, the concept of National Park has been around for over 140 years. The Yellow Stone National Park, USA, the first national park in the world, was officially established in 1872. Although there are different definitions and models of national parks, a general consensus has been reached after over 100 years of development: The difference between national parks and ordinary tourism-oriented parks lies in that the former prioritize conservation in addition to public use. National parks have specific elements and cannot be judged by the names. China's national parks contain the following four elements in line with *Guidelines for Applying Protected Area Management Categories* published by IUCN and China's current situation: 1. Resource elements. A national park should have typical ecological systems and cultural and natural landscapes. It should have a big area and high ecological quality so that ecological processes within it can be sustained. 2. Functional elements. A national park should reserve certain areas to provide opportunities for public recreation, tourism, learning and education. 3. National elements. The central government should play an important and even leading role in managing and financially supporting national parks. 4. Procedural elements. Establishing a national park should follow legitimate procedures, that is, competent departments should designate certain areas as national parks in line with certain procedures and implement planning and management.

Since 2006, some pilot programs on national parks have been carried out in Yunnan, Heilongjiang, Zhejiang and other provinces. The practices are of great significance to integrating protected area management conducted by different departments, implementing effective systems of protected areas and protecting the natural and ecological system with tremendous ecological value. However, developing tourism can bring enormous economic benefits and the general public have a strong demand for educational and recreational opportunities created by protected areas. In this context, local governments may pay more attention to the brand effect and economic benefit the national parks can generate and thus respond actively to pilot programs on national parks. If they just want to create more revenues by utilizing natural resources under the pretext of establishing national parks, it is inevitable that environment and ecology will be damaged. Therefore, it will be difficult to reflect the true characteristics of national parks.

In order to correct the chaotic pilot practices and build an orderly national park system, the central government carried out unified arrangement in 2015. Coordinated by National Development and Reform Commission (NDRC), 13 government ministries jointly issued a pilot scheme and selected nine pilot regions. Pilot work started in three national parks respectively in Sanjiangyuan (Three River Source Region) of Qinghai Province, Kaihua of Zhejiang Province and Shenlongjia of Hubei Province. As the first pilot project, Sanjiangyuan National Park tries to break through the institutional barriers, resolve the issue of fragmented regulation and law enforcement, build an ecological protection management system with clear ownership, well-defined power and responsibilities and effective regulation and set a benchmark for the construction of other national parks in China.

Ecological Governance: A Symphony Played by Different Stakeholders

Promoting the modernization of ecological governance and strengthening the construction of ecological civilization is the only way to alleviating the current severe ecological crisis. It is also a strategic choice for realizing the sustainable development of the country and the Chinese nation. The establishment of the national park system will promote ecological governance modernization and boost ecological progress in terms of awareness raising, institutional arrangement and action.

Regarding awareness raising, the success of the national park system can considerably enhance the sense of national identity and pride, build consensus on ecological protection and achieve synergy in promoting ecological progress. If there is no common belief in ecological protection, ecological governance cannot be supported by all the stakeholders in the society, let alone its modernization. Practices to build the national park system provide an effective recipe for reaching such a consensus. National parks are not only a type of important protected areas, but also represent the typical natural landscape and cultural heritage of a country. People can raise awareness of ecological protection and enhance the concept of environmental protection through appreciating the beautiful natural scenery and splendid cultural heritage so that they can love nature and the country more deeply and gradually understand that protecting national parks can benefit future generations. The enhanced awareness of ecological protection can become a powerful spiritual force in promoting ecological governance modernization.

Regarding institutional arrangement, building a scientific, standardized and efficient protected area system with national parks playing a leading role can promote the modernization of ecological governance. On the one hand, improving the protected area system is the typical content of ecological governance. Establishing the national park system does not mean just setting up only one type of protected area like national parks. Rather, the national park system is used to resolve the issue of fragmented ecological system on the basis of integrating different types of fragmented and highly overlapping protected areas by taking into account the integrality of nature. On the other hand, the establishment of the national park system can enable us to do further research on the key issues of ecological governance modernization. Determining the property right of natural resources within the national parks can provide experience in deepening reform on property right of natural resource assets and unifying management of mountains, waters, forest, lakes and farmland. Clarifying the boundary between conservation and development within national parks can help us to set the upper limit for resource consumption in a scientific way, draw a redline for ecological protection, strictly observe the bottom line for environmental quality and ensure national ecological security. Designing a comprehensive, scientific, reasonable and well-regulated system of protected areas with national parks at the core which features clear division of responsibilities, rights and interests, stable investment and multi-stakeholder coordination will be instrumental in achieving ecological governance modernization.

In practice, many different stakeholders are involved in developing the national park system, which is typical example of transformation from ecological "management" to "governance". Successful national park governance practices in different counties indicate that coordinated actions should be taken by State, enterprises and the public in the whole dynamic process of planning, establishing and maintaining national parks. The central government should conduct top level design of the national park system and cover the main operational expenses. The local governments should implement supporting mechanisms and provide support in management, maintenance and other specific issues. Enterprises and other social organization can provide financial and intellectual support for national parks through franchises, donations and educational and training activities. The public including local residents in particular not only have the opportunity to become staff or volunteers in national parks, but also can make economic contributions through visiting the parks and buying tickets. Diversified stakeholders' engagement in ecological governance is the important guarantee in achieving national ecological governance modernization.

Establishing a National Park System Based on the Rule of Law

The main reason why it has been difficult to integrate the system of protected areas for many years is that the legal framework for the protection is department-based and fragmented. Although several regulations, departmental rules and normative documents such as *Regulation on Natural Protected Areas, Regulation on Scenic Spot Management, Administrative Measures on Forest Parks, Management, Administrative Measures on National Key Parks*, etc. have been issued so far, each of them only covers one single type of protective area with low legal forces. In practice, each department in charge acts in its own way and there is no coordination, so it is difficult to establish an integrated a national park system. Therefore, we should reach consensus on the core content of national park legislation, enact a specific national park law through analyzing and summarizing the difficulties and experiences in the national park pilot programs and bring the establishment of the national park system into the track of the rule of law.

Law on National Parks should include five aspects in general: 1. What is the purpose of the law? The law should clearly define the principle of conservation coming before utilization so as to avoid ambiguity. 2. What is a national park? Based on the IUCN concept, the legal definition should take into consideration the distribution, area and function of China's natural and cultural resources and national and procedural factors. 3. What should be included in the law? The general principles should be conservation first, ensuring public benefits, state playing a leading role and scientific guidance. 4. How is the law used? The applicable scope should include the establishment, planning, conservation, management, utilization, maintenance and monitoring of national parks and other activities. 5. Who is responsible for national parks? This involves the rights and obligations of different stakeholders. The responsibilities of relevant organizations should be clearly defined regarding institutional arrangement, funding, supervision and other aspects. Local residents and the general public should have the right to have access to national parks for tourism and recreation. They should also shoulder the obligations of maintaining the natural state of national parks.

In addition, the law should clearly stipulate such important contents on national parks as the establishment standard, resource ownership, planning and report, management mechanism, funding, supervision and monitoring and assessment mechanism, etc. These main mechanisms should be implemented through making detailed regulations, implementation rules or management or technical standards. The ecological integrity and national representation of national parks should be taken into consideration. The ownership of national park resources should be clarified to ensure national parks get stable financial input to achieve integrated, efficient and unified conservation and management so that the national park system can be established under a comprehensive, systematic and detailed set of laws and regulations.

With the rule of law functioning as a baton, the state, enterprises, the public and other stakeholders will together blow the bugle of the national park system so as to play a harmonious symphony of ecological governance modernization.

第十七章 草原生态治理实现增收增绿

草原是陆地覆盖面积最大的生态系统，被称为"地球的皮肤"，在防风固沙、涵养水源、保持水土、净化空气以及维护生物多样性等方面具有不可替代的作用。我国是世界第二草原大国，拥有各类草原面积近60亿亩，约占国土面积的40%。草原既是我国江河的源头和涵养区、维护生物多样性的"基因库"，也是畜牧业发展的重要资源和牧区群众最基础的生产生活资料。合理开发草原资源、加强草原生态保护对维护国家生态安全和食物安全、促进农牧民增收脱贫、维护民族团结和边疆稳定具有重要战略意义。

生态治理绿了草原富了牧民

近年来，我国草原生态文明建设加快推进，草原生态系统保护与修复成效显著，全国草原生态总体向好，局部地区生态环境明显改善。

第一，草原生态保护优先理念不断深化。2011年，国务院印发《牧区文件》，扭转了过去片面强调草原生产功能、忽视草原生态功能的观念，确定了"生产生态有机结合、生态优先"的基本方针，推动草原生态步入良性循环轨道。《牧区文件》提出把保护基本草原和保护耕地放在同等重要的位置，根据不同牧区资源禀赋条件，因地制宜，科学确定草原保护建设利用的重点。青藏高原中西部、新疆帕米尔高原和准噶尔盆地、河西走廊、内蒙古西部等地区，坚持生态保护为主，实施全面禁牧；青藏高原东部、内蒙古中部、新疆天山南北坡、黄土高原等地区，坚持生态优先，保护和利用并重，严格以水定草、以草定畜，适度发展草原畜牧业；内蒙古东部、东北三省西部、河北坝上、新疆伊犁和阿勒泰山地等地区，坚持保护、建设和利用并重，加大建设力度，全面推行休牧和划区轮牧，实现草畜平衡。

第二，各项政策、工程的实施使草原得以休养生息，全国草原生态总体向好,"风吹草低见牛羊"的美景在一些牧区重现。自 2011 年实施草原补奖政策 5 年来,中央财政累计投入 773.6 亿元,支持实施草原禁牧面积 12.33 亿亩、草畜平衡面积 26.05 亿亩、牧草良种补贴面积 1.2 亿亩。草原生态环境加快恢复,全国重点天然草原牲畜超载率从 2010 年的 30%,下降到 2015 年的 13.5%,基本达到草畜平衡。

实施天然草原退牧还草工程实施以来,工程区草原得以休养生息,草原植被覆盖度和牧草产量明显提高,草群中优良牧草比例逐步增加,草原生态环境逐渐好转。据农业部监测,2015 年退牧还草工程区草原植被覆盖度较非工程区高出 9 个百分点,牧草高度高出 48%,鲜草产量高出 40.2%。通过实施游牧民定居工程,游牧民由四季游牧转为冬春定居补饲、夏秋放牧,促使天然草原植被恢复和产草量提高,大量牧民转产转业,有效缓解了天然草场的承载压力。实施农牧交错带已垦草原治理工程,通过建植多年生人工草地,提高了治理区植被覆盖率和饲草生产、储备、利用能力。

第三,草原生态保护和现代畜牧业发展、牧民增收实现双赢。草原畜牧业特色突出,是特色畜产品供给的主渠道。草原地区坚持在保护中发展、在发展中保护。各地着力加强人工草地和牲畜棚圈等基础设施建设,促进草原畜牧业由天然放牧向舍饲、半舍饲转变,初步实现了"禁牧不禁养、减畜不减肉",草原畜牧业发展方式由粗放型向效益型转变。改良牲畜品种,优化畜群结构,推广舍饲圈养,加快周转出栏,发展加工增值,草原畜牧业综合生产能力明显提高,草原生态保护和畜牧业发展、牧民增收实现双赢。草原补奖政策将近 90% 的资金直补到户,增加了牧民的收入,有力促进了生态补偿脱贫。

草原生态治理能力须不断提升

长期以来,由于自然、地理、历史等原因,我国草原生态文明建设保护存在以下困难和问题。

第一,草原资源底数不清。《中华人民共和国草原法》规定建立草原调查统计制度,定期开展草原资源调查统计工作,但该制度至今尚未建立执行。

实践中，不同部门对于草原资源数据口径不一。例如，20世纪80年代第一次全国草地资源调查数据为58.9亿亩，2011年"国土二调"的数据为43.1亿亩，相差15.8亿亩；江西省草原面积，农业部门统计数据为6663.5万亩，"国土二调"数据为455万亩，相差近15倍。当前，生态文明体制改革正在深入推进，对草原生态资源的精准化管理提出了更高的要求，亟须建立统计调查制度，尽快启动第二次全国草地资源调查。

第二，草原监理和技术服务能力滞后。我国草原地区大多地处偏远、交通不便，草原管护任务十分繁重。长期以来，我国草原监督管理层级低、管理人员少，开展精细化管理难度大。不少地方草原监理机构不健全，全国县级以上草原执法监督机构共900多个，从事草原监理和草原技术推广工作的专业人员各约1万人，平均每60万亩草原仅拥有1名监理人员和1名技术推广人员。由于对草原监理和草原技术服务机构缺乏投入，加之牧区地方财政困难、执法装备短缺落后、技术服务半径小等问题比较突出，制约了草原执法监督和技术推广工作的正常开展。

第三，草原生态系统整体仍较脆弱。我国草原生态总体向好，但仍处于起步恢复阶段。全国中度和重度退化草原面积占到总草原面积的三分之一以上。草原资源环境承载压力较大，大部分牧民收入主要依赖牲畜养殖，加之草原旱灾、鼠虫害和毒害草灾害频发，工业化、城镇化进程中对草原的侵占破坏难以避免，即使是已经恢复的草原生态系统整体仍较脆弱。

第四，畜牧业发展基础薄弱。近年来，国家加大了对草原生态保护的投入，但对牧区牲畜棚圈、青贮窖池、储草棚库等生产性设施的投入少，"水、草、料、林、机"配套水平低，制约了畜牧业发展，在一定程度上影响了牧民实施禁牧和草畜平衡、发展标准化养殖的积极性，不利于草原生态保护成果的巩固和扩大。

着力推进草原生态文明建设

应坚持保护为先、预防为主、制度管控和底线思维，兼顾保障草原生态安全、实现草畜平衡和农牧民增收，着力推进草原生态治理体系与治理能力现代化。

第一，制度创新引领草原生态治理。

1. 建立健全草原法律法规体系。推进《草原法》修订工作，增设草原资源资产产权和用途管制、草原资源资产离任审计、草原资源损害责任追究和草原生态补偿等制度。大幅提高破坏草原资源、生态环境的违法成本，解决对违法处罚偏轻、法律依据不充分等问题。全面修改完善《草原法》配套法规。按照《国家生态安全政策》中"研究制定基本草原保护条例等法律法规"的工作要求，加快《基本草原保护条例》立法进程，使基本草原划定管理工作有法可依。

2. 完善草原休养生息制度和草原保护体系。加快建立以基本草原保护制度、草原承包经营制度、禁牧休牧划区轮牧制度、草畜平衡制度、草原监测评价考核制度为主体的草原休养生息制度体系。加快建立基本草原保护制度，划定和保护基本草原；稳定和完善草原承包经营制度，实现承包地块、面积、合同、证书"四到户"，规范承包经营权流转；全面落实禁牧休牧轮牧和草畜平衡制度，切实减轻天然草原承载压力，实现草原休养生息和永续利用。

3. 提高草原执法监督能力。加快推进草原行政执法体制改革，建立一支权责统一、权威高效的草原行政执法队伍，大幅提升执法能力。加快建立完善草原行政执法与刑事司法衔接机制。实行草原专职管护员制度，专门开展管护工作。配齐配强必要的草原生态巡护技术装备及交通设备。加强业务培训，提高人员综合素质。

4. 加强草原生态监测能力建设。启动全国草地资源调查专项，全面摸清当前草原资源家底，为建立更加严格的资源管控制度、扎实推进草原生态文明建设奠定坚实基础。

第二，多措并举保护恢复草原生态环境，实现"人—畜—草"平衡。

1. 传承优秀草原游牧文化，发展新时代草原文化，树立并践行尊重自然、爱护自然的生态文明理念，加快推进草原生态文明建设。

2. 推进草原生态红线划定工作，严守草原生态空间。制订草原自然保护区规划,加快草原自然保护区建设步伐。

3.坚持工程措施与自然修复相结合、重点突破与面上治理相结合,着力保护和恢复草原生态系统,不断提高草原生态产品生产能力。扩大退牧还草工程实施范围,实施新一轮退耕还林还草工程,扩大退耕还草规模。继续实施京津风沙源治理和石漠化综合治理工程,加大草原有害生物和黑土滩治理力度,实施农牧交错带已垦草原治理。

第三,大力发展现代草原畜牧业。

在保护草原生态环境的前提下,加快现代草原畜牧业发展,促进牧民增收和产业精准脱贫。继续开展粮改饲和草牧业试验试点,促进种植结构调整和草畜配套,推进现代饲草料产业体系建设。实施南方现代草地畜牧业推进行动和振兴奶业苜蓿发展行动,扶持草产品和畜产品生产加工营销。落实《牧区文件》要求,启动实施草原畜牧业转型示范工程,加快培育养殖大户、家庭牧场、牧业专业合作组织等新型经营主体,积极引导龙头企业在牧区发展特色养殖基地,提升草原畜牧业的经营管理水平,提高草原利用水平和生产效率,通过生产发展和牧民增收来促进草原生态保护。

Chapter 17
Grassland Ecological Governance Increases Income and Green Space

Known as "the skin of the earth", grassland is the ecological system which covers the largest land area and plays an irreplaceable role in wind prevention and sand fixation, water conservation, water and soil preservation, air purification and biodiversity maintenance, etc. With a grassland area of 6 billion Chinese *mu* (400 million hectares) or 40% of the total area, China is the second largest country in the world. Grassland is the source and conserving area of the rivers in China and a gene pool of maintaining biodiversity. It also provides important resources for the development of animal husbandry industry and basic means of production and livelihood. Reasonable exploration of grassland and enhancing grassland ecological protection is of strategic significance in maintaining the country's ecological security and food security, alleviating poverty by increasing farmers and herdsmen's income and securing national unity and stability of frontier region.

Ecological Governance Makes Grassland Greener and Herdsmen Richer

In recently years, China has accelerated ecological progress in the grassland area. Remarkable achievements have been made in preserving and repairing grassland eco-system. The overall eco-system has been very good with obvious improvement in some places.

Firstly, the concept that protection of eco-system in grassland area should be prioritized has been enhanced constantly. In 2011, the State Council issued a document on the work in pasturing area which has reversed the notion that only emphasized the productive function of the grassland and ignored the protection of grassland ecological functions. The document confirms the basic principle that production should be organically combined with ecology with the latter given a priority, thus ensuring grassland ecology takes the road of virtuous cycle. The document equally emphasizes the protection of grassland and arable land. In line with the different resource endowment in different regions, priorities should be set to make use of and protect the grassland. In such regions as central and western Qinghai-Tibet Plateau, Pamirs Plateau and Junggar Basin in Xinjiang, Hexi Corridor and western Inner Mongolia, etc., we should stick to the principle of protecting the ecology and banning animal husbandry. In areas like eastern Qinghai-Tibet Plateau, central Inner Mongolia, the south and north slopes of Mount Tianshan in Xinjiang and Loess Plateau, we should stick to the principles of prioritizing ecology, attaching equal importance to protection and utilization, growing grass according to the amount of water available, raising livestock in line with the amount of grass and developing animal husbandry in a moderate way. In areas like east Inner Mongolia, west of the three northeastern provinces, Bashang in Hebei Province, Yilin and area around Mount Atlay in Xijiang, etc., we stick to the principle of placing equal emphasis on protection, construction and utilization, enhancing grassland development and fully implementing closed grazing and rotary grazing in different regions to achieve the balance between grassland and livestock.

Secondly, the implementation of policies and projects enables the grasslands to rehabilitate and the overall grassland ecology has turned for the better. Some beautiful scenes described in an ancient Chinese poem that "rippling through the pastures, north winds blow; the grass bends low, the cattle and sheep, to show" are reproduced in some pastoral areas. Since 2011, the Central Government has invested about 77.36 billion RMB to support the prohibition of grazing in 1.233 billion Chinese mu, balance the grassland area and the number of livestock in 2.605 billion mu and subsidize the use of high-quality seeds of forage grass in 120 million mu. The recovery of the grassland ecological environment has been accelerated. The overload rate of key national natural grasslands decreased from 30% in 2010 to 13.5% in 2015 and the balance of grassland area and livestock numbers has been basically achieved.

Since the project of returning grazing areas to natural grasslands was initiated, the grasslands have rehabilitated with an obvious increase in vegetation coverage and forage yields. High-quality grass accounts for an increasingly big proportion in the grassland and the ecological environment has improved. According to the Ministry of Agriculture, vegetation coverage, grass height and fresh grass output in project areas in 2015 are respectively 9%, 48% and 40.2% higher than those in non-project areas. Through implementing the project of nomadic herdsmen settlement, the lifestyle of the herdsmen has changed from nomadism all year around to grazing in summer and winter and settling down in winter and spring, which has promoted the recovery of natural vegetation and the increase of grass output. A large number of herdsmen have shifted to other industries, effectively alleviating the bearing pressure of natural grassland. The project of returning cultivated land to grassland in the transition zone between cropping areas and grassland has been implemented to improve vegetation coverage and enhance the capability for forage grass production, preservation and use by planting perennial artificial grass.

Thirdly, a win-win situation has been achieved in protecting grassland ecology, developing modern animal husbandry and increasing herdsmen's income. With distinctive characteristics, grassland animal husbandry is the main source of special livestock products. The grassland region should adhere to the principle of attaching equal importance to development and ecological protection. Construction of infrastructure including artificial grassland and animal pens has been strengthened to promote the transformation of grassland animal husbandry industry from natural grazing to drylot feeding or semi-drylot feeding so as to ban grazing, but not to ban raising animals and reduce number of livestock, but not to reduce supply of meat. The development mode of grassland animal husbandry has changed from an extensive pattern to an efficient pattern. We should improve livestock species, optimize livestock structure, promote drylot feeding and livestock rearing in pens and develop value-added processing industries so as to markedly increase comprehensive production capabilities of grassland animal husbandry. A win-win situation has been achieved in protecting the grassland ecology, developing animal husband and increasing herdsmen's income. 90% of the government subsidies for protecting the grassland has been distributed to households, which has increased the herdsmen's income and promoted poverty alleviation through ecological compensation.

Improving Grassland Ecological Governance Capabilities Continuously

For a long time, there have been many problems and difficulties in building China's grassland ecological civilization due to natural, geographical and historical factors.

Firstly, we are not clear about our grassland resources. *Grassland Law* stipulates that a grassland survey and statistics system should be established to conduct regular surverys on grassland resources. However, the system has not been set up yet. In practice, different departments have different data regarding grassland resources. For example, the total area of grassland was 5.89 billion Chinese mu according to the first national grassland resource survey in 1980s. According to the second land resource survey in 2011, the figure was 4.31 billion mu with a gap of 1.58 billion mu. The area of grassland resource in Jiangxi Province is 66.635 million mu according to agricultural department, yet the figure was 4.55 million mu according to the national survey with a gap of over 15 times. At present, the deepening of ecological civilization puts forward a higher requirement on the accurate management of grassland resources. A survey system needs to be set up urgently and the second national survey of grassland resources should be initiated as soon as possible.

Secondly, capabilities for supervising and managing grassland and technical services have lagged behind. Most of China's grassland is located in remote places with difficult transportation, so the task of managing grassland is arduous. For a long time, it has been difficult to conduct fine management of grassland because of shortage of staff and low level of management. There are few regulatory agencies in some places and there are only 900 grassland law enforcement agencies above county level in China with 10,000 staff responsible for grassland management and 10,000 staff for grassland technical support, which means there is only one management staff and one technician every 600,000 mu of grassland on average. Because of lack of input and financial difficulties in these pasturing areas, law enforcement equipments are backward and the technical service radius is so mall, which has restricted the normal operation of grassland law enforcement and technical promotion.

Thirdly, grassland ecosystem is still fragile as a whole. The overall grassland ecology in China is quite good, but it is still in the initial stage of recovery. Area of grassland with moderate or severe degradation accounts for one third of the total grassland area in China. The grassland environment has a large carrying pressure. Most of the herdsmen's income depends on raising livestock. In addition, frequent draught, pests, rats, hazardous grass and the encroachment of grassland due to urbanization have caused inevitable damages, so even if the grassland ecological system has been restored, it still remains fragile.

Fourthly, China has a weak foundation in animal husbandry. In recent years, China has increased the input in protecting grassland ecology. However, input into production facilities like livestock pens and sties, silage silos and grass storage warehouses, etc. has been inadequate and the supporting mechanism for water, grass, forage, forest and machines is not well functioning. These factors have restricted the development of animal husbandry and to a certain extent, have dampened herdsmen's enthusiasm in developing standardized animal husbandry to achieve grass-livestock balance, which is unfavorable to consolidating and expanding the achievements made in grassland ecological protection.

Sparing No Efforts to Promote Grassland Ecological Civilization

We should adhere to the principle of protection first, prevention foremost, rule-based control and management and bottom line-oriented thinking and attach equal importance to ensuring grassland ecological security, achieving grass-livestock balance and increasing farmers and herdsmen's income.

Firstly, grassland ecological governance should be guided by institutional innovation.

1. We should construct a legal framework on grassland. *Grassland Law* should be revised to include property rights of grassland resources and assets and use control, grassland resources and assets audit for outgoing officials, accountability for grassland resource damage and grassland ecological compensation and other mechanisms. We should increase the cost for damaging grassland resources and ecological environment by redressing the issues of light punishment for violating the law and inadequate legal basis. We should comprehensively amend and perfect supporting laws and regulations for *Grassland Law*. According to the work requirement on formulating laws and regulations on protecting grassland set in *National Policy on Ecological Security*, we should speed up the legislative process of drafting *Regulation on Protecting Grassland* so that grassland management can be conducted in line with law.

2. We should improve grassland recuperation system and grassland protection system. We should accelerate the establishment of a grassland recuperation system consisting of basis grassland protection, grassland contractual management and grassland monitoring and evaluation, etc. Grazing on certain areas of grassland will be banned or temporarily suspended, rotational paddock grazing will be introduced, and efforts will be made to strike a balance between grass and livestock. We should establish a system of protecting basic grassland as soon as possible. The system by which collective grassland is contracted out for operation to individual households will be kept stable and improved to ensure that the plot and area of every piece of grassland contracted out is measured accurately, contracts are signed, and contracting certificates are granted. Proper procedures will be introduced for the transfer of grassland under such contracts. We should fully implement the system of banning, suspending or rotating grazing and grass-livestock balance so as to reduce the carrying pressure of natural grassland and achieve grassland recuperation and sustainable development.

3. We should improve law enforcement and supervision capability. We should speed up reform on grassland administrative law enforcement system by establishing an efficient and authoritative law enforcement team integrating power and responsibilities and enhancing law enforcement capability. The administrative

law enforcement system should be integrated with criminal justice system. Full-time rangers should be employed to take care of the grassland. They should be equipped with necessary technical equipment and vehicles for patrolling the grassland. Training should be conducted to improve their overall quality.

4. We should enhance capabilities for monitoring grassland ecology. We should kick off special surveys on national grassland resources and understand the comprehensive picture of current grassland resources, which can lay a solid foundation for establishing a stricter resource control and management system and promoting grassland ecological civilization.

Secondly, various measures should be taken to protect and restore grassland ecological environment to achieve the balance between people, livestock and grass.

1. We should inherit splendid grassland nomadic culture, develop grassland culture in the new era, establish and practice the concept of ecological civilization which respects and loves nature and promote grassland ecological progress.

2. We should establish a grassland ecological redline and strictly preserve grassland ecological space. We should make plans on protecting natural reserves on grassland and speed up the construction of these reserves.

3. We should strive to protect and restore grassland ecological system by combining engineering measures with natural restoration and by launching key projects and conducting overall improvement. We should continuously increase productivity of grassland ecological products. We should expand the scope of returning the grazing land to grass and carry out a new round of projects of returning cultivated land to forest and grass. We will press ahead with the project of controlling the source of sandstorms affecting Beijing and Tianjin and comprehensively addressing the expansion of stony deserts. We should step up our efforts to control grassland pests and black soil land and restore the cultivated land to grassland in transition areas from agriculture to animal husbandry.

Thirdly, we should vigorously develop modern grassland animal husbandry industry.

We should accelerate the development of modern grassland animal husbandry industry to increase herdsmen's income and alleviate poverty under the precondition of protecting grassland ecological environment. We should carry out pilot projects of replacing grain crop cultivation with feed crop cultivation and grass-based animal husbandry to promote planting structure adjustment and the balance between grass and livestock and facilitate the development of modern forage industry. We should develop modern grassland animal husbandry, revitalize the development of alfalfa for dairy industry in South China and support the processing and marketing of grass products and livestock products. We should implement the requirement set in *Document on the Pasturing Area,* kick off some demo projects on transforming grassland animal husbandry industry and fostering some big animal husbandry businesses, family ranches and animal husbandry cooperatives and other new type of business entities. We should actively guide leading companies to develop special husbandry bases, improve operational and management level of grassland animal husbandry, increase grassland utilization rate and production efficiency and promote grassland ecological protection through increasing production and farmers' income.

第十八章　恢复湿地以养护"地球之肾"

湿地是地球自然生态系统的重要组成部分，其生态服务价值在各类生态系统中居首位，具有涵养水源、净化水质、调节局地气候、蓄洪防旱、维持生物多样性等多种生态功能，被誉为"地球之肾"。湿地是淡水的"蓄水池"，维持着约2.7万亿吨淡水，保存了全国96%的可利用淡水资源；湿地作为"物种基因库"，还是众多植物、动物特别是水禽生长的乐园和候鸟迁徙的"中转站"；湿地还是重要的"储碳库"和"吸碳器"，对于固碳释氧、应对气候变化具有重要意义。

保护湿地换得绿水长流

我国湿地资源丰富，据2014年发布的第二次全国湿地资源调查结果显示，全国湿地总面积5360.26万公顷，湿地率5.58%，另有水稻田面积3005.70万公顷。只有不断扩大湿地面积、维护湿地生态功能，才能碧波荡漾、绿水长流，实现人与自然的和谐。

自1992年加入《关于特别是作为水禽栖息地的国际重要湿地公约》（以下简称《湿地公约》）以来，我国高度重视湿地的保护与合理开发利用，通过制定并完善湿地保护法律制度、实施湿地修复工程、开展湿地资源调查监测、提高公众湿地保护意识等措施积极扩大湿地面积、修复湿地功能。截至2015年年底，我国已有国际重要湿地49处、湿地自然保护区600多个、湿地公园1000多个，湿地保护率达44.6%，以湿地自然保护区为主体，国际重要湿地、湿地公园、湿地保护小区等并存的湿地保护体系基本形成。

保护湿地是维护国家生态安全，保障经济社会可持续发展和建设生态文明的需要。作为滨海湿地面积最大的省份，江苏省着力保护湿地、修复生态环境，为每年过境中转的逾300万只鸟类打造安全舒适的"驿站"。重庆市为守住三峡库区的绿水青山,强化消落带生态的修复与治理,试验出"池杉+中华蚊母+卡开芦+牛鞭草"等5个消落带类型湿地乔灌草相结合的植被修复构建模式。我国的商品粮基地大都分布在湿地比较集中的区域。如果湿地消失,这些商品粮基地将最终丧失生产能力。建立商品粮基地周边的湿地保护体系,事关重大。黑龙江省已在三江平原地区形成了较为完善的湿地保护网络,保护总面积达60.7万公顷,同时恢复湿地8000公顷,为保障黑土地大粮仓粮食稳产高产发挥了重要作用。为有效解决湿地资源权属不清、权责不明等问题,2015年国家提出在甘肃、宁夏两省区先行开展湿地产权确权试点,为建立完善湿地资源保护管理制度奠定坚实基础。2016年12月12日国务院办公厅印发《湿地保护修复制度方案》,明确建立一系列湿地保护修复制度,对新形势下湿地保护修复做出了部署安排。

莫让"地球之肾"衰竭

虽然我国湿地保护在过去一段时期取得了显著成效。但是湿地保护工作仍然面临巨大挑战。长期以来,在工业化、城镇化进程中过度的开发利用方式只顾局部、短期利益,单纯追求经济增长,正在把湿地一块一块地"填平烤干"。污染、围垦、基建占用、过度捕捞和采集以及外来物种入侵造成湿地面积萎缩、功能退化。这种对湿地重开发、轻保护,重索取、轻投入的开发利用方式致使湿地不堪重负,已超出生态系统的承载力。

湿地保护治理体系和治理能力不足、科技支撑不强、全社会湿地保护意识薄弱使得湿地保护任务更具复杂性和艰巨性。目前,我国在森林、草原、海洋、湿地等自然生态系统中,唯独湿地缺少国家层面的立法,其他相关法律法规多是从水、土壤、生物等单一要素角度进行规范。虽然已有20多个省份制定了湿地保护条例,但全国范围内的湿地保护法律法规尚未出台,使各级主管部门保护湿地法律依据不足,无法有效约束破坏湿地的行为。同时,湿地实行"综合协调与分部门管理相结合"的管理体制,涉及林业、国土、水利、环保、农业、海洋等多个部门,部门之间的权责划分不明确,综合协调不到位,容易出现"多龙治水"的情形。加之国家对湿地管理资金投入不足,导致基层执法队伍薄弱、设施落后。

湿地是鸟类的天堂、物种的基因库。长期的不合理开发利用方式和治理能力的不足致使湿地面积不断减少、功能退化、生物多样性面临巨大威胁,直接破坏了各类生物的栖息环境,严重影响着生态安全。第二次全国湿地资源调查结果显示:全国湿地总面积5360.26万公顷,与第一次调查(1995~2003)同口径比较,湿地面积减少了339.63万公顷,接近于海南岛的面积。1949年以来,我国滨海湿地累计减少219万公顷,占全国滨海湿地总面积的50%。据调查,从2004~2014年,全国湿地鸟类种类呈现严重减少趋势。青海湖环湖区开垦面积约20万平方公顷左右,脊椎动物减少了34种,斑头雁仅10年就减少1200只。沿海滩涂湿地是许多候鸟漫长迁徙路程的唯一"落脚点"。由于滩涂湿地面积的减少,候鸟迁徙途中无法补充体能,导致鸟类体质下降,影响到鸟儿的越冬和繁衍活动。长江流域水鸟栖息地的减少和破坏直接威胁着众多稀有鸟类的生存,中华鲟、江豚、白头鹤等珍稀濒危物种种群数量不断下降;洞庭湖鱼类的种数从114种减少到80种。在湿地减少的面积中,沼泽湿地减少最多,湖泊湿地减少次之。沼泽湿地相对于其他类型湿地,具有更加复杂的生物链网络,对于维持生态系统功能具有更高的价值,因此沼泽湿地的减少,更令人担忧。四川若尔盖高原沼泽区自20世纪50年代起开始排干沼泽辟为牧场,目前已排水疏干沼泽20余万公顷。随着沼泽湿地疏干排水,地表无积水,局部出现盐渍化,附着在地表的植被无法生长,沙化土地面积急剧扩大。

全面推进湿地保护与修复

要实现到2020年全国湿地面积不低于8亿亩,湿地生态功能总体稳定的目标,必须积极扩大湿地面积、科学修复退化湿地、提高湿地生态功能,将湿地保护作为推进生态文明建设的重要抓手,努力完善湿地保护治理体系,不断提升治理能力。

完善湿地治理法律制度体系。加快推进湿地立法。尽快出台《湿地保护条例》，协调建立湿地执法协作机制，严厉查处违法利用湿地的行为，使湿地保护工作有法可依。划定落实湿地保护"红线"。确保红线区湿地面积不减少、性质不改变、功能不退化。规范湿地用途管理制度，合理设立湿地相关资源利用的强度和时限。将湿地保护专项规划纳入水资源管理、流域综合管理、土地利用等多个重大规划之中，协调布局、统筹管理。将湿地面积、湿地保护率、湿地生态状况等指标纳入各级政府评价考核范围。完善湿地分级管理体系。根据生态区位、生态系统功能和生物多样性的重要性，将湿地列入不同级别的名录，定期更新。建立湿地资源资产管理体制，探索编制湿地资源资产负债表。推进湿地生态效益补偿制度。建立中央财政湿地保护补助专项资金，依据湿地价值评估，对湿地占用进行经济补偿。

积极扩大湿地面积。完善湿地保护空间规划。构建完善的湿地保护体系，不断提高湿地保护率。重点加强对国际和国家重要湿地、湿地自然保护区、国家湿地公园等重点湿地的保护与监管。按国家公园的理念管理湿地生态系统，全面提升湿地管理合力。推进退耕还湿，鼓励农民进行"水改旱"，让湿地的生态效益尽快凸显出来。强化湿地保护管理的科技支撑，提高科技对湿地保护的贡献率。例如研究我国湿地保护区成效定量化评价方法，基于最新湿地自然保护区评价数据库和湿地遥感数据等资料，对保护成效进行全面的定量化评估。

科学修复退化湿地，提升湿地生态系统的整体功能。大力实施湿地生态修复工程，提升湿地生态系统功能。编制湿地保护修复工程规划，对集中连片、破碎化严重、功能退化的沼泽、河流、湖泊、滨海湿地进行修复和综合整治。培育湿地生态产业，建立湿地恢复的社区参与机制。发展湿地花卉苗木、湿地旅游等湿地生态产业，提升湿地经济效益，构建湿地社区共管共建共享机制。着眼流域尺度恢复湿地生态系统。本着共建"山水林田湖"生命共同体的理念，维护流域湿地生态系统的完整性与功能的多样性，而不仅局限于湿地生态系统本身。

Chapter 18
Restore Wetland to Protect "Kidney of the Earth"

Wetland is an important part of earth's natural ecosystem. Its ecological service value holds the first place among all kinds of ecosystems. It has various ecological functions, such as water source conservation, water purification, regulating local climate, flood storage and drought-resisting, and maintaining biological diversity, etc. It is honored as the "kidney of the earth". Wetland is the "impounding reservoir" of fresh water, which maintains about 2.7 trillion tons of fresh water and keeps 96% available fresh water resources of the country. As "species gene reserve", wetland is the paradise for the growth of various plants, animals, especially waterfowls, and "transfer station" for bird migration. In the meantime, wetland is the important "carbon storage" and "carbon absorption device" with the significance for fixing carbon, releasing oxygen, and dealing with climate change.

Protecting Wetland for a Better Ecology

Our country has abundant wetland resources. According to the second national wetland resource investigation result which was announced in 2014, the total wetland area of our country was 53,602,600 hectares; the rate of wetland was 5.58%. In addition, the paddy field area was 30,057,000 hectares. Only by expanding wetland area continuously and maintaining ecological function of wetland can we get green rippling lake, forever green water and realize harmony between human and nature.

Since joining *Convention on Wetlands of International Importance Especially as Waterfowl Habitat,(Wetland Convention)* in 1992, our country has been paying great attention to the protection, reasonable development and utilization of wetland. Through taking the measures of making and completing legal system of wetland protection, implementing wetland restoration project, carrying out wetland resource investigation and monitoring, and improving wetland conservation awareness of the public, we expand wetland area actively and restore wetland function. Up to the end of 2015, our country has 49 international important wetlands, more than 600 wetland natural reserves, and more than 1,000 wetland parks. The wetland protection rate reaches to 44.6%. The wetland protection system has been formed by wetland natural reserve as subject, and coexistence of international important wetland, wetland park and wetland conservation community.

Wetland protection is the demand for maintaining national ecological safety, ensuring sustainable economic and social development, and building ecological civilization. As the biggest province of coastal wetland area, Jiangsu Province focuses on protecting wetland and restoring ecological environment. It builds safe and comfortable "station" for more than 3 million birds that pass here for migration every year. In order to keep green hills and clear water of the Three Gorges Reservoir Region, Chongqing strengthens the restoration and treatment of fluctuating belt ecology, and obtains 5 fluctuating belt type vegetation restoration construction modes with the combination of wetlands, trees, brushes and grasses through tests, such as the mode of "pond cypress + distylium + phragmites karka + hemarthria". Most of the commodity grain bases of our country are distributed in the regions with concentrated wetlands. If the wetlands disappear, these commodity grain bases would lose production capability finally. It is important to set up wetland protection system around the commodity grain bases. Heilongjiang Province has formed relatively complete wetland protection network in Sanjiang Plain region. The total protection area reaches to 607,000 hectares. In the meantime, it restores 8,000 hectares wetlands, and exerts important function for ensuring high and stable yield of grain of black soil large granary. In order to solve the problem of obscure ownership, unclear duty and right of wetland resource efficiently, China came up with the policy of using Gansu Province and Ningxia Province as experimental units for carrying out confirmation of wetland ownership first in 2015. It lays solid foundation for building and completing wetland resource protection management system. On December 12, 2016, the General Office of the State Council released *Plan of Wetland Protection and Restoration System* which established a series of wetland protection and restoration systems and made arrangement for wetland protection and restoration under new situation.

Never Exhausting the Kidney of the Earth

Although China made remarkable progress on wetland protection in the past period, it still faces great challenge for wetland protection work. During the process of industrialization and urbanization for a long time, the wetlands are being "filled up and dried" piece by piece due to the excessive development and utilization, and the simple pursue of economic growth by only taking partial and short-term interests into consideration. The wetland area is shrunk, while its functions are degenerated by pollution, inning, infrastructure occupation, overfishing and excessive collection, and invasion of alien species. This kind of development and utilization way which emphasizes development but neglects protection; emphasizes acquisition but neglects investment has made the wetland overwhelmed. It has exceeded the bearing capacity of the ecological system.

Due to the deficient wetland protection and management system, insufficient governance capabilities, weak science and technology support, weak wetland protection awareness of the whole society, wetland protection is complicated and difficult. At present, our country has legislation at the country level on the natural ecosystems of forest, grassland, ocean, etc., but not on wetland. Most of the other related laws and regulations specify the requirement from single element aspect of water, soil and biology, etc. Although there are more than 20 provinces which have made wetland protection regulation, nationwide laws and regulations on wetland protection have not been issued. As a result, the competent departments at all levels do not have sufficient legal basis of wetland protection. They are not able to restrain the behavior of damaging wetland effectively. In the meantime, it carries out the management system which "combines comprehensive coordination and separation departmental management", involving forestry, territory, water conservancy, environmental protection, agriculture, ocean and many departments. The right and duty among the departments are not divided clearly; while the comprehensive coordination is not sufficient. It easily leads to the situation of "coexistence of multiple departments in water environment regulation". Furthermore, the country does not invest enough capital on wetland management which leads to the weak grassroots law enforcement team and backward facilities.

Wetland is the paradise of birds and gene pool of species. Due to long-term unreasonable development and utilization and insufficient governance capabilities, it decreases wetland area and declines the functions of the wetland. The biological diversity faces greatest challenge. It damages habitat environment of various organisms directly and influences ecological safety seriously. According to the result of the second national wetland resource investigation, the total wetland area of the country was 53,602,600 hectares. Comparing with the first investigation (1995～2003) under same caliber, the wetland area decreased 3,396,300 hectares which is close to the area of Hainan Island. Since the 1949, the costal wetland of our country has decreased 2,190,000 hectares accumulatively. It occupies 50% of total costal wetland area. According to investigation, the bird species of national wetland decreased seriously from 2004 to 2014. The cultivated area surrounding the Qinghai Lake was about 200,000 hectares. The vertebrates decreased 34 species. Bar-headed goose decreased by 1,200 only in ten years. Costal beach wetland is the only "foothold" for many migratory birds during the long migration path. Due to the decrease of costal wetland, the migratory bird cannot replenish its body during the migration. It declines the physical deterioration of the birds and influences the activity of birds for living through the winter and reproduction. The decrease and damage of waterfowl habitat in Yangtze River threaten the survival of various rare birds directly. The population quantity of Chinese sturgeon, finless porpoise, white-head crane, and other rare and endangered species is decreasing continuously. The species of fishes in the Dongting Lake decreases from 114 to 80. Among the decreased wetland area, the largest area of reduction is swamp wetland; while the lake wetland takes the second place. Comparing with the other types of wetland, swamp wetland has more complicated biologic chain network. It has higher value on maintaining the function of ecological system. Therefore, people are more concerned about the decrease of swamp wetland. The swamp area of Zoige Plateau in Sichuan Province has started to drain the swamp for pasture since the 50s of last century. At present, it has drained the swamp for more than 200,000 hectares. Along with the drainage and dewatering of swamp wetland, there is no water on the land surface; salinization occurs in some areas; the vegetation which attaches to the land surface is not able to grow; and the desertification land area extends greatly.

Promoting Wetland Protection and Restoration Comprehensively

In order to realize the target of no less than 800 million mu (1 mu=0.0667 hectares) nationwide wetland area and overall stability of wetland ecological function in 2020, it must expand wetland area actively; restore degraded wetland scientifically; improve ecological function of wetland; treat wetland protection as a significant aspect of promoting ecological civilization construction; strive to complete wetland protection and management system; and improve the governance capabilities continuously.

Complete the legal system of wetland management; accelerate wetland legislation; issue Wetland Protection Regulation as soon as possible; coordinate and establish wetland law enforcement cooperation mechanism; investigate and punish the behavior of illegal use of wetlands strictly; make the wetland protection work with law for basis; delimit and implement the "red line" of wetland protection; ensure the wetland in red line region without area decrease, property change, and function degeneration. Regulate management system of wetland application; set the strength and time limit of related wetland resource utilization reasonably. Include the specialized planning of wetland protection into water source management, integrated watershed management, land utilization and many major plans. Make coordinated layout and overall management. Include the indexes of wetland area, wetland protection rate, wetland ecological situation into assessment scope of governments at all levels. Complete wetland classification management system. According to the importance of ecological niche, ecosystem function, and biological diversity, list wetlands into the lists with different levels and update them periodically. Build wetland resource and assets management system; explore and compile wetland resource and assets balance sheet; promote wetland ecological benefit compensation system; establish special subsidies of wetland protection from central finance; and make economic compensation for wetland occupation in accordance to wetland value evaluation.

Expand wetland area actively; complete wetland protection space plan; establish complete wetland protection system; and improve wetland protection rate continuously; emphasize and strengthen the protection and supervision of international and national important wetland, wetland nature reserve, national wetland park, and other important wetlands. According to the concept of national park, manage wetland ecological system, and improve wetland management comprehensively. Promote the progress from cultivated land to wetland and encourage the farmer to transform rice fields into dry land so as to highlight ecological benefits of wetland as soon as possible. Strengthen science and technology support of wetland protection management, improve contribution rate of science and technology on wetland protection. For example, make research on effective quantitative evaluation method of wetland reserve in our country, and conduct comprehensive quantitative evaluation on the protection effect on the base of latest wetland nature reserve evaluation database and wetland remote sensing data.

Restore degenerated wetland scientifically; improve overall function of wetland ecological system; vigorously implement wetland ecological restoration project; improve wetland ecological system function. Compile wetland protection restoration project plan; make restoration and comprehensive regulation on concentrated and contiguous swamp, river, lake and costal wetland with serious fragmentation and functional degeneration. Cultivate wetland ecological industry; build community participation mechanism of wetland restoration. Develop wetland ecological industry of wetland flowers and plants and wetland tourism; improve wetland economic benefit; establish joint administration, establishment and sharing mechanism of wetland community; focus on watershed scale and restore wetland ecological system. By insisting the concept of building the life community of "hill, water, forest, farm and lake", maintain the completeness of watershed wetland ecosystem and functional diversity rather than restricting to wetland ecosystem only.

第十九章　抚育经营森林　换得山绿民富

森林是"地球之肺",是陆地生态系统的主体,是发展现代林业的资源基础,是人类生存发展的重要生态保障。据2014年发布的第八次全国森林资源清查数据,全国森林面积2.08亿公顷,森林覆盖率21.63%,森林蓄积151.37亿立方米。稳步扩大森林面积,提升森林质量,增强森林生态功能是发挥林业多种功能、多重效益的根本保证。

可持续经营森林确保青山常在

保护生态、改善民生是林业最基本、最重要、最核心的任务和职责。只有解决民生问题才能缓解生态问题,生态建设的好坏也会直接影响民生的改善。

经营森林稳固国土生态安全屏障。我国森林资源不但数量持续增加,为现代林业发展提供资源,而且质量稳步提升、效能不断增强,源源不断生产出丰富的生态产品,保障着国家生态安全。据联合国粮农组织发布的《2015年全球森林资源评估报告》,我国已成为全球年度森林面积增长最快、森林蓄积稳定增长的国家。在自然林持续增长的基础上,人工林快速发展;森林质量不断提高,森林功能不断由"木材供应基地"向"生态安全屏障"转变。据测算,我国森林生态系统每年提供的主要生态服务价值达12.68万亿元。随着森林资源的增长和质量的提高,森林植被生物量、涵养水源量、森林碳汇能力等进一步增强。

发展现代林业惠民富民。我国林业产业规模持续扩大,林产品供给能力不断提升,林产品贸易跃居世界首位。林业产业总产值从2001年的4090亿元增加到2015年的5.94万亿元,15年增长了13.5倍,对7亿多农村人口脱贫致富做出了重大贡献。通过大力发展花卉苗木、木本油料、特色经济林、林下经济、森林旅游、森林康养等特色产业，在劳务分配、产业开发、科技服务、金融支持等方面向贫困群众倾斜，精准到人、到户，林业发展同时助力精准扶贫和绿色发展。

林业改革既保护生态，也保障民生，将"绿水青山"变成"金山银山"。国有林场和国有林区是国家最重要的生态安全屏障。2011年，河北、浙江等7省被列为国有林场改革试点省。2014年，重点国有林区停止天然林商业性采伐试点启动。国有林场和国有林区试点工作进展顺利，取得了显著成效。2015年，中共中央、国务院印发《国有林场改革方案》和《国有林区改革指导意见》，全面启动国有林场和国有林区改革。自2008年集体林权制度改革以来，全国已确权面积27.05亿亩，让1亿多农户直接受益，林权保护管理体系日益完善，实现了"山定权、树定根、人定心"，走出了一条"林农得实惠、企业得资源、国家得生态"的生态富民新路。

美丽中国不能缺林少绿

当前，我国林业发展还面临着巨大的压力和挑战。森林资源总量仍然相对不足，同时质量欠佳、分布不均；森林覆盖率远低于世界平均水平，木材对外依存度接近50%；缺林少绿、生态脆弱的现状迫切需要改变。

实现森林增长目标任务艰巨。在2005年的基础上要实现到2020年森林面积增加4000万公顷、森林蓄积量增加13亿立方米，2030年森林蓄积量增加45亿立方米的目标任务艰巨。从第八次全国森林资源清查结果看，森林"双增"目标前一阶段完成良好，森林蓄积量增长2020年目标已完成，森林面积增加目标已完成近六成，但森林面积增速开始放缓。现有未成林造林地面积减少，同时宜林地质量下降，且2/3分布在西北、西南地区，立地条件差，加之传统的廉价劳动力、土地投入要素等优势逐步丧失，造林难度增大，如期实现森林面积增长目标任务十分艰巨。

森林"提质"迫在眉睫。森林质量不高是我国林业最突出的问题。我国林地生产力低，每公顷蓄积量只有世界平均水平的2/3，影响了生态综合功能的发挥。近几年，我国在植树造林方面花费了大量的人力、物力、财力，造就了世界第一的造林面积。然而生态环境质量并未像预期好转。全国森林质量参差不齐，甚至有1/4的森林处于亚健康和不健康状态。此外，林业发展方式落后、经营管理粗放也是导致森林质量低、造林存活率和成林率低的重要原因。

林业治理能力有待提高。森林治理体系的理念、许多制度已不适应新形势下林业发展的要求，长期以来，林业科技支撑薄弱、信息化建设滞后、生产机械化程度低、人才队伍薄弱，营林、培育公益林等资金存在巨大缺口。我国相对集中连片的林区多位于老少边穷等地区，林区道路、供电、饮水、通信等基础设施建设和公共事业长期落后。

多措并举推进生态林业、民生林业

必须树立绿色发展的理念、全面深化林业改革、创新林业治理体系，采取扩面、增绿、提质、增效等多种措施全面提高生态林业和民生林业发展水平，保护好每一寸绿色，真正将"绿水青山"变成"金山银山"。

推进国土绿化"种好树"。开展大规模国土绿化行动，加强天然林资源保护、三北防护林、京津风沙源治理等重点生态工程建设，扩大新一轮退耕退牧、还林还草规模，集中连片建设森林，着力增加生态脆弱区林草植被。坚持封山育林、人工造林并举，乔灌草、带片网相结合，自然修复与人工修复并重。深入开展全民义务植树活动，创新国土绿化投入机制，吸引社会力量参与造林绿化。充分发挥国有林区、国有林场在国土绿化中的带动作用。积极开展碳汇造林，探索推进林业碳汇交易。开展森林城市建设，构建完备的城市森林生态系统，为城市群、城市、村镇可持续发展提供坚实的生态保障。加大能源林基地建设，规模化培育能源林。

提升森林治理能力"护好树"。完善森林法律法规体系，加快修改《中华人民共和国森林法》（以下简称《森林法》）。本着有利于保护发展和合理经营利用森林资源的原则，进一步明确《森林法》的调整范围，适度放宽林业管制,突出分类经营、生态优先等原则,推动生态文明建设。建立健全森林资源资产产权制度,加强对林权流转交易的监督管理。完善林业分类经营体制改革,对公益林和商品林实施不同的管理、经营措施。积极稳妥地推进重点国有林区改革,健全国有林区经营管理体制。积极推进国有林场改革,按照公益事业单位管理要求,明确国有林场生态公益功能定位。深化集体林权制度改革,改革和创新集体林采伐管理、资源保护、生态补偿、税费管理等相关政策。推动建立全球森林治理体系,为全球绿色经济和可持续发展做出更大贡献。

科学经营森林"管好树"。提高森林质量,关键在于加强森林经营。实施森林质量精准提升工程。调整森林结构,加强森林抚育,构建健康稳定优质高效的森林生态系统。坚持保护优先、自然修复为主,坚持数量质量并重、质量优先,坚持封山育林、人工造林并举。完善天然林保护制度,促进天然林资源休养生息,全面停止天然林商业性采伐。因林施策,针对不同类型、不同发育阶段、不同演替进程的森林,采取更加精准的经营技术措施;公益林严格保育、兼用林多功能经营、商品林集约经营。建立森林经营规划制度。把森林质量提升作为约束性指标纳入政府目标考核体系。推进政府与社会资本共同筹资的多元投入机制。全面提高造林质量,优化森林树种结构,不断提升森林的多种功能效益。完善相关技术标准规程,强化森林资源保护。

发展林业绿色经济"用好树"。破解高效培育技术瓶颈,实现技术创新,提高林木材积生长量。提升林产加工业,推进木材储备基地建设,不断增强木材和林产品的有效供给能力。推进油茶、核桃等木本粮油产业发展。建设林木种苗、花卉、竹藤、生物药材、木本调料等生产基地。大力发展林下经济,增加生态资源和林地产出。打造特色林业产品示范园区。优化人造板、家具、木浆造纸、林业装备制造业等产业布局。做大做强森林旅游、森林康养等林业服务业。

Chapter 19
Nurture and Manage Forest to Make Mountains Green and People Rich

Forest is the "lung of the earth". It is the subject of terrestrial ecosystem, resource base for developing modern forestry, and important ecological guarantee of human existence and development. According to the data of the eighth national forest resource inventory which was released in 2014, the forest area of the country was 208 million hectares; the forest coverage rate was 21.63%; the forest stock volume was 15.137 billion cubic metres. The measures of expanding forest area stably, improving forest quality, and strengthening forest ecological function are the fundamental guarantees for exerting multiple functions and benefits of forestry.

Sustainable Forest Management Ensuring Green Mountains Forever

Ecological protection and improvement of people's livelihood are the most basic, important and core task and responsibility for forestry. Only the problem of people's livelihood is solved, the ecological problem could be mitigated. The good and bad ecological construction will also influence the improvement of people's livelihood directly.

Manage forest and stabilize homeland ecological security barrier. The forest resource of our country increases continuously on the quantity and provides resource to the development of modern forestry; in the meantime, its quality improves stably with enhanced efficiency. It produces abundant ecological product continuously and ensures the national ecological security. According to *2015 Global Forest Resource Assessment Report* by United Nations Food and Agricultural Organization, our country has become the country with the fastest annual forest area growth and stable growth of forest stock in the world. On the base of sustainable growth of natural forest, man-made forest develops rapidly. The forest quality improves continuously. The forest function transfers from "timber supply base" to "ecological safety barrier" continuously. According to estimation, the main ecological service value which is provided by the forest ecological system in China every year reaches RMB 12.68 trillion Yuan. Along with the growth of forest resource and quality improvement, the biomass of forest vegetation, water conservation quantity and forest carbon sequestration capacity are further improved.

The development of modern forestry benefits and enriches people. The forestry industry scale of our country expands continuously; while the supply capacity of forest products also improves continuously. Our country has taken the first place in the world for the forest products. The total output value of forestry industry increased from RMB 409 billion Yuan in 2001 to RMB 5.94 trillion in 2015. It increased 13.5 times in 15 years and made great contribution to overcome poverty and achieve prosperity for more than 700 million rural population. Through vigorous development of flowers and plants, woody oil plants, characteristic economic forest, under-forest economy, forest tourism, forest conservation, and other characteristic industry, incline to poor people on the aspect of labor distribution, industry development, science and technology service, and financial support with the accuracy to individual and household; develop forestry and assist targeted poverty alleviation and green development at the same time.

Forestry reform not only protects ecological environment, but also ensures people's livelihood. It changes "green hills and clear water" to "mountains of gold and silver". State-owned forest farm and forest area are the most important ecological safety barrier of the country. In 2011, 7 provinces, such as Hebei, Zhejiang, etc. were listed as reform pilot provinces of state-owned forest farm. In 2014, the pilot project of stopping commercial logging of natural forest in the key state-owned forest area was started. The pilot work in state-owned forest farm and forest area went well and obtained remarkable effect. In 2015, the Central Committee of the Communist Party of China, and the State Council printed and distributed *Reform Plan of State-owned Forest Farm*, and *Reform Guidelines of State-owned Forest Farm*, and started the reform of state-owned forest farm and forest area comprehensively. Since the reform of collective forest right system in 2008, it has confirmed the right of area for 2.705 billion mu (1 mu=0.0667 hectares) in China; benefited more than 100 million peasant households directly; improved forest rights protection management system day by day; realized the work of "determining the right of hill, fixing the root of tree, and stabilizing heart of people"; and created a new ecological road of "benefits to forest worker, resource to enterprise, and ecological environment to country" which enriches people.

Green Is Indispensable for Beautiful China

At present, the forestry development of our country still faces huge pressure and challenge. The total forest resources are relatively insufficient; meanwhile, the quality is poor with uneven distribution. The forest coverage rate is far from world average level. The foreign-trade dependence of wood is close to 50%. The current situation of lacking forest and green and weak ecological environment needs to be changed urgently.

The task of realizing the target of forest growth is difficult. It is arduous task for realizing the target of increasing 40 million hectares forest area and 1.3 billion cubic meters forest stock volume by 2020; increasing 4.5 billion cubic meters forest stock volume by 2030 on the base of 2005. According to the result of the eighth national forest resource inventory, the "double-increase" target of forest in the previous stage completed well. The target of increasing forest stock volume by 2020 has been completed. The target of increasing forest area has been finished for 60%. However, the growth speed of forest area has slowed down. The existing immature forest land area decreases; meanwhile, the quality of suitable land for forest declines. Two thirds of the lands are distributed in northwest and southwest regions with poor site conditions. Furthermore, the advantages of traditional cheap labor force and land input elements are losing gradually which increase the difficulty of building forest. It is arduous task for realizing the target of forest area growth as scheduled.

"Improve the quality" of forest is extremely urgent. Low forest quality is the most outstanding problem of forestry in China. The forest land productivity of our country is low. The forest stock volume per hectare is only 2/3 of world average level. It influences the development of comprehensive ecological function. In recent years, our country has spent a large number of manpower, material resources and financial resources on forestation and achieved the first forestation area in the world. However, the quality of ecological environment does not turn better as expected. The forest quality in the country is on various levels; even 1/4 forests are in the state of sub-health and ill-health. In addition, backward forestry development mode and extensive operation and management are the important reasons which cause low forest quality, low forestation survival rate, and low rate of forest forming.

The governance capabilities of forestry shall be improved. The concept of forest management system and many systems are not applicable for the demand of forestry development under new trend. There are problems of weak science and technology support on forestry, backward information construction, low production mechanization degree, weak talent team, huge funding gap on forest culture and management, and cultivating public welfare forests for a long time. Most of the relatively concentrated and contiguous forest regions in our country are located at poor and remote areas where the infrastructures and public utilities of road, power supply, drinking water and communication in the forest regions lag behind for a long time.

Taking Multiple Measures to Promote Ecological Forestry and Improve People's Livelihood

It is a must to set up the concept of green development; deepen forestry reform comprehensively; bring forth new ideas of forestry management system; adopt multiple measures of expanding area, increasing green, improving quality, and increasing effect for improving the development level of ecological forestry and forestry of people's livelihood comprehensively, protecting the green by each inch, and changing "green hills and clear water" to "mountains of gold and silver" in deed.

Promote land greening, "plant the trees well". Carry out large scale action of land greening; strengthen the protection of natural forest resources, and key ecological project constructions, such as Three North Shelterbelt, and Beijing-Tianjin Sandstorm Source Control, etc.; expand the new round of scale for returning cultivated land and grazing land to forest and grassland; build concentrated and contiguous forest; focus on increasing vegetation cover in ecological fragile area. Insist on closing hillsides to facilitate forestation and man-made forest simultaneously; combine trees, brushes, and grasses with belt, piece and net; pay equal attention to natural restoration and human-induced restoration. Carry out national voluntary tree-planting activities deeply; bring forth new ideas for land greening investment mechanism; attract social force for participating in forestation and land greening; exert the leading function of state-owned forest area and forest farm in land greening. Carry out carbon sequestration forestation actively; explore and push forestry carbon sequestration transaction. Carry out forest city construction; build complete urban forest ecosystem; provide solid ecological guarantee for sustainable development of urban agglomeration, city, villages and small towns. Expand energy forest base construction, and cultivate large scale energy forest.

Improve forest governance capabilities, "protect the trees well". Complete forest laws and regulation system; and accelerate to revise *Forest Law*. Based on the principle of benefiting to the protection, development, reasonable management and utilization of forest resources, further confirm the adjustment scope of *Forest Law*; relax forestry regulation moderately; highlight the principle of classified management and ecological priority; push ecological civilization construction. Establish sound forest resource and assets property right system; strengthen the supervision and management on the transaction of forest right transfer. Complete the reform of forestry classification management system; undertake different management and operation measures for public welfare forest and commercial forest. Promote the reform of key state-owned forest region actively and steadily; complete operation and management system of state-owned forest region. Push state-owned forest farm reform actively; confirm the positioning of ecological public welfare function of state-owned forest farm in accordance to the management requirement by public service unit. Deepen the reform of collective forest property right system; reform and innovate collective forestry cutting management, resource protection, ecological compensation, tax management and other related policies. Push and establish global forest management system and make more contribution to global green economy and sustainable development.

Scientific forest management, "manage the trees well". The key of forest quality improvement is on strengthening forest management. Carry out forest quality precision upgrading project. Adjust forest structure; strengthen forest nurture; establish healthy, stable, high quality and efficient forest ecosystem. Insist on giving priority to protection and natural restoration; insist on paying attention to quantity and quality simultaneously with quality priority; insist on closing hillsides to facilitate forestation and man-made forest simultaneously. Complete natural forest protection system; facilitate rehabilitation of natural forest resource; stop commercial cutting of natural forest entirely. Implement policy in accordance to feature of forest; for the forest with different type, developmental stage and progression of succession, adopt more accurate technical management measures; protect and nurture public welfare forest strictly, undertake multi-function management for multi-functional forest, make intensive operation for commercial forest. Set up forest operation planning system. Regard forest quality improvement as obligatory target and include it into target evaluation system of the government. Push diversified investing mechanism which is co-financed by the government and social capital. Improve forestation quality comprehensively; optimize tree species structure of forest; improve multiple functional benefits of the forest continuously; complete related technical standard regulation; and strengthen forest resource protection.

Develop green economy of forestry, "use the trees well". Solve the technical bottleneck of efficient cultivation; realize technology innovation; improve forest volume increment. Upgrade forest products processing industry; push the construction of timber reserve base; strengthen effective supply capability of timber and forest product continuously. Push the wood, food and oil industry development, such as tea-oil tree, juglans regia, etc. Build the production base of forest seedling, flowers, bamboo and rattan, biological medicine, and woody spices, etc. Make a great effort to develop under-forest economy, increase ecological resource and forest land output. Build characteristic demonstration garden of forestry product. Optimize the industry layout of artificial board, furniture, wood based paper, and forestry equipment manufacturing, etc. Make the forestry service industry bigger and stronger, such as forest tourism, forest rehabilitation, etc.

第二十章　创新沙漠治理　培育金色产业

我国是世界上荒漠化面积最大的国家，同时也是全球治沙取得重大成果的实践典范。根据《第五次中国荒漠化和沙化状况公报》，截至2014年，全国荒漠化土地面积261.16万平方公里，占国土面积的27.20%；沙化土地面积172.12万平方公里，占国土面积的17.93%。治理沙漠对于改善生态环境、实施精准扶贫、推进生态文明建设具有重大的现实意义。

从"沙进人退"到"人沙和谐"

荒漠化和沙化土地呈面积减少、程度减轻趋势。经过长期不懈的努力，三北防护林、京津风沙源治理、退耕还林、退牧还草等重点生态工程持续取得进展。目前我国荒漠化和沙化状况呈现整体遏制、持续缩减、功能增强、成效显著的良好态势。根据2015年发布的第五次全国荒漠化和沙化监测结果显示，我国荒漠化和沙化土地面积自2004年以来连续10余年保持"双缩减"，其中荒漠化土地面积已从20世纪末年均扩展1.04万平方公里转变为目前的年均缩减2424平方公里，沙化土地面积由20世纪末年均扩展3436平方公里变为目前的年均缩减1980平方公里。

发展沙产业使沙害变沙利，黄沙变黄金，沙漠成良田。1984年钱学森提出，具有充沛阳光资源的沙漠、戈壁可以发展成为农业型产业空间。应该大力发展"多采光、少用水、新技术、高效益"的沙产业。30多年来，在沙产业理论的指导下，在我国西部沙区已初步形成了以灌草饲料、中药材、经济林果、沙漠旅游等为重点的沙区特色产业，开发出了饲料、药品、保健品等一大批沙产品，并带动了种植、加工、物流等相关产业的发展，沙产业链不断延长，产值不断增加，成长起了一批从沙漠中淘金的龙头企业，昔日的不毛之地如今生机盎然。沙产业的发展使沙害变沙利，黄沙变黄金，沙漠成良田，增加了土地存量，为粮食安全奠定了基础。

沙漠治理仍然任重而道远

沙漠治理认识不高、资金不足、知识不够。总体上看，我国各地沙漠治理工作仍不平衡，一些地方存有畏难情绪和消极思想，资金投入力度不够、管理不善，"生态欠账"问题突出，甚至有些地方存在故意污染沙漠的情形。目前，国内外治理沙漠主要通过围栏封育、退耕还林还草、植树造林等生物措施增加植被覆盖度或通过工程和化学措施固定沙质地表，减少风沙活动。这些方式在短期内产生了一定的生态效益，但忽视了生态系统整体的功能恢复和协调发展。同时，荒漠化治理的物理、化学措施由于忽视了沙区的社会经济效益，其可持续性也受到质疑。

沙产业生产规模小，组织化程度低，整体效益不显著。许多荒漠化、沙化地区生态虽然有所好转，但农牧民却并未脱贫致富，存在"生态强、经济弱"的现象。防沙治沙项目投融资模式简单、资金压力大，只依靠国家拨款投资，引入市场机制和民间资本力度不够。沙产业是知识密集的高科技农业型产业，项目周期长、投资风险大、企业规模小、科技创新力度不足、资源综合利用水平不高、产业精深加工能力弱等因素制约了沙产业的进一步发展。

共创防沙、治沙、用沙新篇章

树立绿色发展理念，强化法治保障，推广中国经验。防沙治沙不是只有一种方式，而应树立绿色发展理念，因地制宜，针对不同情况提出技术、产品、治理一体化的解决方案。逐步完善沙漠土地"三权分置"制度设计，延长沙漠土地承包权、经营权流转年限，为沙漠治理提供法治保障。清晰沙漠土地产权，依法维护经营主体从事沙漠治理所需的各项权利，使沙漠土地资源得到更有效合理的利用。加强国际合作，推进"一带一路"地区防沙治沙，共筑丝绸之路经济带生态屏障；推广沙漠治理"中国经验"，共促世界各国生态治理。

　　创新发展沙漠经济，打造金色产业。要改变单纯以固沙为主要目标的传统防沙治沙理念，坚持"防沙治沙与用沙"相结合。充分开发沙漠潜能，发挥沙区光热资源充足、土地资源广阔的优势，采取"多采光、少用水、新技术、高效益"的技术路线，积极发展绿色富民产业，培育沙产业龙头企业，提升沙产业的质量和效益。发展沙漠经济，将沙漠治理与沙区植物、新能源和旅游等资源产业化结合起来，依靠产业创新、技术创新、金融创新等手段，构建以"科技支撑—生态修复—产业发展—社会进步"为主线的沙漠治理新型模式。

　　"沙漠就是资源，生态就是生意"，沙区扶贫是全国扶贫攻坚难啃的硬骨头。治理沙漠必须与发展经济、改善民生、精准扶贫有效结合起来，让沙区群众从治沙中受益，看到治沙的经济、生态前景。充分考虑沙区的自然条件、社会环境，因地制宜利用沙区资源禀赋，探索出一条把"治沙"和"治穷"相结合的生态扶贫道路。

　　统筹节水、集水、护水、调水手段，合理、高效利用水资源。让"沙漠变绿洲"，水资源是决定性因素，要推广节水灌溉技术，提高水资源利用效率。选育和引进节水高效的牧草、农作物、林果品种，实现"高投入—高产出—高收益"的良性循环。慎重开采地下水，保证合理水位，避免过度开采地下水使浅层水位下降，造成土地沙化。充分开发利用"边缘水资源"，包括天然降水的收集，沙漠居民的废水回收利用和沙漠地下咸水的挖掘及淡水转化。保护管理水资源，防止水质变化。对沙漠及周边的河流、湖泊和水库进行保护。积极调入水资源，研究论证南水北调西线工程。

Chapter 20
Innovate Desert Governance and Cultivate Golden Industry

China is the country with the largest desertification area in the world. In the meantime, China is also a practice model of achieving significant results on desertification control in the world. According to The fifth *Bulletin of Desertification and Sandification Status in China*, up to 2014, the national desertification land area was 2,611,600 square kilometers which occupied 27.20% of the national territorial area; the sandification land area was 1,721,200 square kilometers which occupied 17.93% of the national territorial area. Desert governance has great realistic significance for improving ecological environment, implementing targeted poverty alleviation, and promoting ecological civilization construction.

From "Desert Advancing and Duman Retreating" to "Harmony Between Human and Desert"

There is a trend that the area of desertification and sandification decreases; and the severity degree is alleviated. Through long-term and unremitting efforts, the key ecological projects, such as Three North Shelterbelt, and Beijing-Tianjin Sandstorm Source Control, returning cultivated land to forest, returning grazing land to grassland, have obtained continuous progress. At present, the desertification and sandification situation of our country appears the good situation of overall containment, continuous shrinkage, and remarkable effectiveness. According to the result of the fifth national desertification and sandification monitoring in 2015, the land area of desertification and sandification of our country has been keeping "double shrinkage" for more than 10 consecutive years since 2004. The desertification land area has changed from annual average expansion of 10,400 square kilometers at the end of last century to annual average shrinkage of 2,424 square kilometers now; the sandification land area has changed from annual average expansion of 3,436 square kilometers at the end of last century to annual average shrinkage of 1,980 square kilometers now.

Develop desert industry; change the desert hazard to desert benefit; change yellow desert to gold; change desert to fertile farmland. In 1984, Mr. QIAN Xuesen, a leading Chinese scientist, came up with the opinion that the desert and gobi with abundant sunshine resources could be developed into agricultural industrial space. It shall make great effort to develop desert industry of "more lighting, less water, new technology, and high benefit". For more than 30 years, under the guidance of desert industry theory, we have formed preliminary desert land characteristic industry in desert land of western China which focuses on shrub grass fodder, traditional Chinese medical herbs, economic forest and fruit tree, desert tourism, etc.; developed a batch of desert products, such as fodder, medicine and health care products; and driven the development of planting, processing, logistics and related industries; extended the desert industry chain and increased production value continuously; developed a batch of leading enterprises which seek gold in desert. The land which did not grow anything in the past has been transferred to the land which is full of vitality. The development of desert industry changes the desert hazard to desert benefit; changes the desert to fertile farmland; increases land inventory; and lays foundation for food security.

Desert Governance: An Arduous Task

There are problems of low awareness, insufficient fund and knowledge for desert governance. In general, the desert governance work throughout China is not balanced; the people of some regions have the fear of difficulty and negative thoughts; the capital investment strength is not enough; improper management; outstanding problems of "ecological debt"; even some regions pollute the desert on purpose. At present, the desert governance at home and abroad mainly takes the biological control measures through fencing enclosure, returning cultivated land to forest and grassland, and forestation for increasing vegetation coverage; or takes engineering and chemical measures for fixing desert surface and reducing wind-sand activity. These methods have certain ecological effect in short period. However, they neglect the overall function restoration of ecosystem and harmonious development. In the meantime, because the physical and chemical measures of desertification governance neglect the social and economic benefits of the desert area, so the sustainability is questioned.

There are problems of small scale production of desert industry, low organization degree, and unremarkable overall benefit. Although the ecological environment turns better in many regions of desertification and sandification, the peasant and herdsman could not overcome poverty and achieve prosperity. The phenomenon of "strong ecological environment, but weak economy" still exists. For the desert prevention and control project, the financing and investment mode is simple with great capital pressure. It only relies on the fund investment by the government now. The strength for bringing in market mechanism and private capital is not enough. Desert industry is a knowledge-intensive and high-tech agricultural industry. The factors, such as long project cycle, big investment risk, small scale enterprise, deficient science and technology innovation strength, low comprehensive resource utilization level, and weak industrial intensive processing capability, etc., restrain the further development of the desert industry.

Preventing, Controlling and Utilizing Sand

Establish concept on green development, strengthen legal guarantee and promote Chinese experience. It is not just one way for desert prevention and governance. We shall set up green development concept; adjust measures to local conditions; come up with integrated solution of technology, product and governance for different situation. Complete the "Three Rights Separation" system design of desert land gradually; extend the transferring time limit of contracting right and operational right of desert land; and provide legal guarantee for desert government. Clarify desert land property right; protect all the required rights of operating entity for engaging in desert governance by law; and make more efficient and reasonable utilization of desert land resource. Strengthen international cooperation; push desert prevention and governance in "Tne Belt and Road Initiative" regions; build ecological barrier of Silk Road economic belt together; promote "Chinese experience" in desert governance; and push ecological governance in various countries in the world together.

Develop desert economy in innovative way and create gold industry. It shall change the traditional desert prevention and governance concept which only regards stabilization of desert as main target; insist the combination of "desert prevention, governance and utilization". Fully develop the potential of desert; exert the advantage of desert area of sufficient sunshine and heat resources and wide land resource; adopt the technical path of "more lighting, less water, new technology, and high benefit"; develop green industry which enriches people actively; cultivate leading enterprise of desert industry; improve the quality and benefit of desert industry. Develop desert economy; combine desert governance with resource industrialization of plant, new energy and tourism in desert area; build new mode of desert governance with the main line of "science and technology support-ecological restoration-industry development-social progress" by the means of industrial innovation, technical innovation and financial innovation.

"Desert is resource, ecological environment is business". Poverty alleviation in desert area is a tough nut to crack in national poverty alleviation. It must combine desert governance with economy development, improvement of people's livelihood, targeted poverty alleviation effectively; benefit the people in desert area during desert governance; and make people see the economic and ecological prospect of desert governance. Take full consideration of natural condition and social environment in desert area; adjust measures to local conditions and utilize resource endowment in desert area; explore an ecological and poverty alleviation road with the combination of "desert governance" and "poverty governance".

Make overall planning of water saving, water collection, water protection and water adjustment; use water resource reasonably and effectively. Water resource is a decisive factor for "changing desert to oasis". It shall promote water-saving irrigation technique and improve utilization efficiency of water resource. Select breeding and bring in water-saving efficient grass, crop and fruit variety; realize virtuous circle of "high investment-high output-high benefit". Consider carefully for mining groundwater; ensure reasonable water level; prevent the dropping of shallow water level and sandification of land due to excessive exploitation of groundwater. Fully develop and utilize "marginal water resources", including the collection natural rainfall, reclamation of waste water, excavate saline groundwater of desert and transformation of fresh water. Protect and manage water resource; prevent water quality change. Protect the desert, and surrounding river, lake and water reservoir. Transfer water resource actively; research and demonstrate western route of water diversion from South to North.

第二十一章　推进海洋生态治理　守护"蓝色家园"

海洋是生命的摇篮，占地球表面积的71%，是资源的"聚宝盆"、污染的"消纳场"、气候的"调节器"。我国是海洋大国，大陆海岸线长达1.8万公里，管辖海域面积约300余万平方公里，面积500平方米以上的海岛6900余个。我国海洋生物多样性十分丰富，分布有滨海湿地、海岛、海湾、入海河口等多种类型的海洋生态系统。然而，我国海洋生态环境保护现状不容乐观，近海污染和近岸生态破坏问题突出。

优化海洋空间开发保护格局，推动海洋产业结构绿色转型，发展可持续的蓝色经济，维护海洋生物多样性、保护海洋生态系统，建成"水清、岸绿、滩净、湾美、物丰"的美丽海洋，是衡量国家综合竞争力的重要指标，是全球海洋治理的长期课题，对于建设"21世纪海上丝绸之路"、推动海洋生态文明建设、实施"海洋强国"战略具有重要意义。

海洋生态治理现代化进程

我国通过制定、修订一系列法律法规、规划纲要方案，提高海洋开发利用的"生态门槛"，探索海洋生态治理体制机制，推进海洋产业绿色转型，选划建立各类海洋保护区，加强海洋生态修复整治，完善海洋生态环境监测网络，不断推进海洋生态治理体系和治理能力现代化。

建立健全海洋生态环境治理体系。自1982年颁布《中华人民共和国海洋环境保护法》（以下简称《海洋环境保护法》）以来，我国已出台《中华人民共和国海域使用管理法》《中华人民共和国海岛保护法》《中华人民共和国深海海底区域资源勘探开发法》《无居民海岛开发利用审批办法》等涉海资源环境保护相关法律法规数十部；特别是十八大以来，《全国海洋主体功能区规划》《国家海洋局海洋生态文明建设实施方案》《海岸线保护与利用管理办法》

《围填海管控办法》《海洋督察方案》等多项政策出台；生态红线制度、重点海域污染物总量控制制度、海洋生态补偿制度、海洋生态损害赔偿制度、资源环境承载能力监测预警制度、区域限批制度等一系列制度陆续实施，共同勾画出"生态+海洋治理"的新模式，使海洋生态环境法治基础不断夯实，治理体系日益健全。

各地在海陆统筹、责任考核、保护示范、整治修复、生态监测等领域积极探索。中央及各地积极推进海陆统筹协调机制，例如2009年国务院同意建立渤海环境保护省部际联席会议制度；浙江由海洋部门牵头负责全省岸线及周边区域的保护与管理，理顺了海岸带这一海陆统筹关键区域的管理职责。福建厦门、山东青岛连续多年探索实施地方政府海洋生态环境保护目标责任考核；辽宁实施"河长制"，通过流域和入海断面考核将减排责任分解到流域内各市县。通过设立海洋自然保护区、海洋特别保护区、海洋公园、海洋生态文明建设示范区等各类示范区、保护区树立治理典范。实施"蓝色海湾""南红北柳""生态岛礁"等重点工程，积极推进海洋生态建设和整治修复，加快"美丽海洋"建设。初步建立起层级分明、覆盖面广的海洋生态环境监测网络，涵盖多个监测类别，实现对管辖海域全覆盖的动态监测。

海洋生态治理面临多重挑战

目前我国海洋生态治理体系缺乏刚性约束，海洋产业发展欠缺绿色规制，海陆分割体制依然存在，海洋生态环境监测能力不足，海洋生态治理面临多重挑战。

海洋产业发展缺乏绿色引领，海洋生态保护亟须刚性约束。进入21世纪以来，我国海洋经济快速发展，目前已占国内生产总值（GDP）近1/10，成为国民经济的重要组成部分。但是，传统海洋产业欠缺绿色规制，发展方式粗放。海上风电、深海养殖、深远海资源开发等新兴海洋产业处于起步阶段，海洋产业亟待转型升级。由于陆源污染、大规模填海造地、过度捕捞以及船舶溢油等因素影响，我国滨海湿地大量丧失，近岸海洋生态系统退化，部分海域严重污染，海洋生态已经基本处于"亚健康"状态。总体上看，当前我国海洋生态治理法律体系还不系统、不完备，缺乏刚性约束，甚至存在诸多空白，依法治海亟待进一步加强。

海陆分割体制依然存在。现有海洋行政体制未充分考虑海洋生态环境的整体性。海洋与环保、公安、法院、监察、检察等部门执法协调联动机制不健全，信息互通不到位；各部门之间海洋生态环境数据不一致、标准不统一等现象频出。海洋污染的绝大多数是陆源污染，而且陆源污染往往是跨行政区域的，污染源的扩散受天气、潮流等因素影响较大，具有不可控性，所以海洋环境污染只看结果，不看源头，单以行政区域的监测结果作为考核指标是片面的。

海洋生态治理能力有待提升。海洋生态环境监测数据分别来自海洋和环保两个部门，而两个部门的监测站位、方法和程序不同，导致监测数据不统一；海洋监测网络规划布局、数据质量管理、标准规范建设、信息集成应用、监测能力建设等方面还存在不足。海洋生态保护经费投入不足，海洋生态补偿资金渠道单一。海洋生态保护与修复技术还存在较大差距，海洋高新技术对海洋经济的贡献率和科技转化率还很低。

多管齐下提升海洋生态治理水平

健全海洋生态治理体系，完善海洋生态法治保障，建立海洋环境保护目标责任制，形成基于生态系统的海洋综合管理体制；着眼海洋经济绿色转型，积极培育绿色、低碳和循环海洋产业；推进海洋生态修复，强化海洋生态监测；强化科技、人才、资金作用；提升全民海洋意识，参与全球海洋治理；全面提升海洋生态治理能力，推进海洋生态文明建设，实现由"管海"走向"治海"。

健全海洋生态治理体系。加强海洋立法工作，健全海洋生态法治保障，系统梳理海洋生态文明建设领域法律法规和标准。将人海和谐、海陆统筹、生态治海的理念纳入海洋事业的顶层设计，使海洋生态文明从制度建设、能力布局到行动设计全面、充分融入海洋治理总体格局。建立健全海洋资源产权法律制度，制定完善海洋空间开发保护、海洋生态环境保护、海域海岛自然资源保护和利用等配套法律法规；加快推进海洋督察、海洋主体功能区、围填海总量控制、自然岸线保有率控制、海洋生态补偿、海域环评限批等重点制度；加快提升海洋资源环境管理的依法行政水平，用最严格的制度和最严密的法治来保护海洋生态环境，全面推进依法治海。加快制定《海洋基本法》，将海洋生态保护作为海洋发展的基本原则。

建立海洋生态环境保护目标责任制，完善相关行政法规与规章，将海洋生态环境保护纳入生态文明建设评价考核的大盘子中予以统筹考核，或单独评价考核，重视考核结果的运用。建立跨行政区和海陆联动机制，明确海洋生态环境保护地方政府负责制和目标责任制。结合生态文明体制改革，优化海洋生态环境保护监管体制，形成统一且有所分工和配合的海洋生态环境保护监管格局，在明确各自分工的基础上，形成海陆统筹的考核体制机制，建立党政领导海洋生态环境损害责任追究制度体系。

加快海洋经济绿色转型。注重绿色引领，追求人与海洋和谐、当代与后代可持续、经济—社会—海洋协调的发展模式。推广生态健康养殖，加快渔业产业结构调整，充分发挥渔业的碳汇功能。加快海洋清洁能源开发，扩大海水利用应用规模。统筹协调海陆发展，把海洋资源的开发利用、保护与陆域经济发展结合，与沿海的产业群、港口群、城市群结合，与推进内陆腹地的开放开发结合，实现海陆之间资源互补、产业互动、布局互联。

有效推进海洋生态修复，强化海洋生态环境监测。推进海洋生态保护和修复，加强海洋生物多样性保护。推进"蓝色海湾"工程，结合陆源污染治理，实施环境综合整治、退堤还海、清淤疏浚等措施，推动海湾整治修复。实施"南红北柳"生态工程，南方种植红树林等，北方种植柽柳等，因地制宜开展滨海湿地、河口湿地生态修复工程。推进海岛保护与开发，促进有居民海岛有序开发，加强无居民海岛保护，并强化特殊用途海岛管理。强化海洋灾害风险防范能力，增强海洋应对气候变化的能力。推动海洋生态环境监测体制改革，优化国家监测布局，加强基层监测机构建设，整合多部门海洋监测力量，构建统一的海洋环境监测网。

强化科技、人才、资金作用，提升全民海洋意识，参与全球海洋治理。加大海洋基础研究和前沿海洋技术开发力度，重点在深水、绿色、安全的海洋高技术领域取得突破。确保"科海协同"工作机制顺利开展，促进科技创新与海洋发展"互通互联"，推进海洋科技创新向创新引领型转变。加快海洋教育发展，发挥多学科协同优势，培养复合型海洋人才。加大对海洋生态保护领域的资金投入。加强海洋生态文明理念宣传，提高全民族海洋意识。建立公共参与机制，营造全社会关爱海洋、保护海洋，支持海洋生态文明建设的良好氛围。拓展双边海洋合作，引导多边区域合作，加快国际海域调查与极地考察能力建设，全面参与国际海洋事务。

Chapter 21
Promote Marine Environmental Governance and Protect the "Blue Home"

The ocean is the cradle of life, accounting for 71% of the surface area of the earth. It is a "cornucopia" filled with resources, a midden heap that digests pollution and a regulator of the climate. China has a large area of ocean, with its coastline stretching up to 18,000 kilometer, its jurisdiction of the sea area of about 300 million square kilometers and more than 6900 islands of over 500 square meters. China boasts rich marine biodiversity and multiple types of marine ecosystems, such as coastal wetlands, islands, bays, estuaries and so on. However, the marine environmental protection in China is far from enough. Offshore pollution and ecological damages are very severe.

The important indicators of a country's comprehensive competitiveness and a protracted subject of global marine environmental governance is the optimization of the development and protection pattern of marine space, transformation of marine industry into a environmental friendly structure, a sustainable blue economy, maintenance of marine biodiversity, protection of marine ecosystem and creation of beautiful marine environment with clear water, green shores, clean beaches, pretty bays and abundant resources. These are of great significance to the construction of the "21st Century Maritime Silk Road", the promotion of marine ecological civilization construction and the implementation of the strategy of marine power.

Modernizing Marine Environmental Governance

Through the development and revision of a series of laws, regulations, plans, guidelines and schemes, China aims at raising the threshold for the exploration of marine resources, developing the mechanism of marine environmental governance, shifting the marine industry into a greener structure, establishing various types of marine protected areas, strengthening restoration of marine eco-system, revamping the marine ecological environment monitoring network and constantly promoting the modernization of the marine environmental governance system and governance capabilities.

We will establish and improve the marine environmental governance system. Since promulgation of the *Marine Environmental Protection Law*, China has enacted the *Law on the Administration of the Use of Sea Area*, the *Island Protection Law*, the *Law on the Exploration and Development of Resources on Deep Seabed Area*, the *Methods of Approving the Development and Exploration on Uninhabited Islands*, all together over 10 laws and regulations on marine resources and environmental protection. In particular since the 18th National Congress of Communist Party of China, the country has issued the *National Major Marine Functional Zone Planning*, the *Marine Ecological Civilization Construction Scheme*, the *Administrative Measures on Coastline Protection and Utilization*, the *Administrative Measures on Reclamation Management*, the *Marine Inspection Scheme* and other policies; implemented the policy of ecological red lines, control of the total amount of marine pollutants, marine ecological compensation, penalties to marine ecological damages, monitoring and early warning of carrying capacity and restriction in certain regions and so on. These constitute to a new model of ecological plus marine governance, consolidating the foundation for the legal system of marine environment and improving the governance system.

Local governments are actively exploring how to coordinate the land and sea management, assess responsibilities, set example for protection of marine environment, remediate the environment and monitor the ecological environment. The central and local governments have actively set up the land-sea coordination mechanism. For example, in 2009 the State Council agreed to build an inter-ministerial joint meeting system for environmental protection of Bohai Sea; in Zhejiang, the Oceanic Administration took the lead to protect and manage the province's coastlines and the surrounding area; in Xiamen of Fujian and Qingdao of Shandong, local governments are assessed according to their realization of marine environmental protection goals for many years; in Liaoning, officials in different cities and counties are put accountable for reduction of waste discharge and assessed based on discharge to the waters and estuaries. Good examples are being set, such as marine nature reserves, marine special protection areas, marine parks, all kinds of demonstration areas including marine ecological civilization demonstration area and protected areas. Key projects, such as, the Blue Bay, the Mangrove in the South and Tamarix Chinensis in the North, the Ecological Island Reef, are implemented to promote marine ecological construction and remediation and speed up the creation of beautiful ocean. A preliminary multi-tiered marine ecological environment monitoring network will be established to conduct real time monitoring of different types in the whole marine jurisdiction.

Multiple Challenges Facing Marine Environmental Governance

At present, China faces multiple challenges in its marine ecological governance system, such as no legally binding restrictions, lack of regulations on the development of marine industry, separation of the land and the sea system and weak capacity of marine ecological environment monitoring.

The development of marine industry needs to be ushered into an environmental friendly era and legally binding restrictions need to be set urgently for the marine ecological protection. Since the beginning of the 21st century, marine economy has grown rapidly, now accounting for nearly 1/10 of gross domestic product (GDP) and making up an important part of the national economy. However, Regulations are still lacking for the green development of the traditional marine industry and the development mode is still extensive. Offshore wind power, deep sea farming, deep sea resources development and other emerging marine industries are still at the initial stage and the marine industry needs to be transformed and upgraded. Due to land-based pollution, large-scale land reclamation, over-fishing and oil spill, coastal wetlands have largely disappeared, coastal marine ecosystems are degraded and some marine areas are seriously polluted in China. Marine ecology has been basically in the "sub-health" state. On the whole, the current legal system of marine ecological governance in China is unsystematic and incomplete, lacks legally binding restrictions and leaves many areas unregulated. Legal grounds are urgently needed for marine governance.

The land and the sea system are still separated. The existing marine administrative system does not fully consider the integrity of the marine ecological environment. The mechanism that coordinates law enforcement between the marine agencies and environmental protection agencies, public security agencies, courts, supervisory agencies, prosecution services and other departments is yet to be improved and information exchanges among these units are insufficient. Marine environmental data and standards are inconsistent among various departments. The vast majority of marine pollution comes from the land, which often knows no administrative areas. The proliferation of pollution sources is highly affected by weather, currents and other factors, leading to great uncertainties, so it is one-sided to just look at the results of one administrative area but ignore the sources of the marine environmental pollution.

The capacity of marine environmental governance is yet to be improved. Marine ecological environment monitoring data are mainly generated by the marine and the environmental protection departments, but their monitoring data are often inconsistent because of different monitoring stations, methods and procedures. There are still deficiencies in the planning and layout of the marine monitoring network, data quality management, standard making, application of information integration and monitoring capacity building and so on. Marine ecological protection funding is insufficient and marine ecological compensation funds still come from a single channel. Marine ecological protection and remediation technology still lags behind and marine high tech is less applied and benefits little to the marine economy.

Multipronged Measures to Improve Marine Environmental Governance

An integrated marine governance mechanism based on the ecosystem will be formed with the building of a comprehensive marine environmental governance system, an improved legal system of the marine ecological protection and the actability system of marine environmental protection. Focus will be given to the transformation of the marine economy into a green, low carbon and circular industry. Technologies, professionals and funds will been given a full play. Awareness will be increased to motivate people to participate in marine environmental governance. We will fully improve the capability of marine environmental governance and promote the construction of marine ecological civilization so as to achieve transformation from marine control to governance.

We will improve the marine environmental governance system. Marine law making will be enhanced to create a comprehensive legal system of marine environmental governance and systematically sort out the marine ecological civilization construction laws. Concepts such as harmony between human and sea, coordination between land and sea system and marine governance in an ecological way will be included into the top-down design of the marine industry so as to integrate the system building, capacity building and action design of marine ecological civilization to the overall landscape of marine governance. We will establish and improve the legal system of maritime resources property rights, develop wide-range supporting laws and regulations related to the marine space development and protection, marine environmental protection, sea island natural resources protection and utilization. Essential measures will be taken, including marine inspection, building of marine main functional areas, the total reclamation control, maintenance of natural shorelines, marine ecological compensation and sea area environmental assessment. Law enforcement in marine governance will roll out in an all round way, including strengthened administrative management, stringent institutions and strict rules. We will speed up the development of the *Basic Law of the Ocean* and make the marine ecological protection a basic principle of marine development.

The accountability system of marine environmental protection targets will be established and relevant administrative measures and regulations will be improved. The work of marine environmental protection will be considered in the assessment of ecological civilization construction, or will be assessed independently. The function of assessment results will be emphasized. The interregional and land-sea coordination mechanism will be set up to clarify the responsibilities and targets of marine environmental protection for local governments. The marine environmental protection and supervision system will be optimized in combination with the reform of ecological civilization system to create a unified regulatory pattern of marine environment protection with clear division of labor. An overall land-sea assessment system will be built based on clear division of labor and the official accountability system for marine ecological damages will also be established.

We will speed up the transformation to green marine economy. We will lead this transformation in pursuit of a sustained development model that promote harmony between man and the ocean and coordination among economic, social, marine development. We will promote healthy ecological farming, speed up the adjustment of fishery industry structure, and give full play to the fishery's role as a carbon sink. We will accelerate the development of marine clean energy and expand the scale of seawater utilization. By combining the development, utilization and protection of marine resources with economic development on the land, coastal industrial clusters, port groups and urban agglomeration, we will promote coordinated development between the land and the sea to to achieve complementary effect of land and sea resources, industry interaction and interconnection of layouts.

We will effectively promote the marine ecological restoration and strengthen monitoring of marine ecological environment. The marine ecological protection and restoration will be promoted to protect marine biodiversity. The Blue Bay project will be carried out to promote remediation of bays in ways of land-based pollution control, comprehensive environmental remediation, embankments removal and dredging. By planting the Mangrove in the South and Tamarix Chinensis in the North, the project of the Mangrove in the South and Tamarix Chinensis will be implemented to remediate coastal wetlands and estuary wetland according to local conditions. We will protect and develop islands by promoting the orderly development of inhabited islands, enhancing protection of uninhabited islands and improving management of islands with special purposes. We will enhance capability of preventing marine disaster risks and responding to climate change. The marine ecological environment monitoring system will be reformed to optimize the national monitoring layout, strengthen the construction of grassroots monitoring institutions, integrate multi-sectoral ocean monitoring and build a unified marine environmental monitoring network.

We will make full use of science and technologies, professionals, funds, improve marine awareness and encourage people to participate in global marine governance. The marine basic research and cutting-edge marine technology development will be increased to make breakthrough in the field of deep water, green, safe marine high-tech. We will ensure mechanism for coordination between the State Oceanic Administration and Ministry of Science and Technology to be carried out smoothly to connect technological innovation and marine development and promote innovation-driven marine development. The development of marine education will be accelerated to play multi-disciplinary synergies and cultivate comprehensive marine professionals. Finical input will be ramped up to marine ecological protection. Publicity of marine ecological civilization will be promoted to improve people's awareness in marine environmental protection. Public participation mechanism will be built to create a climate where the whole society care for and protect the ocean and support the construction of marine ecological civilization. We will expand bilateral maritime cooperation and take a lead in multilateral regional cooperation. We will speed up capacity building of international maritime and polar investigation and fully participate in international ocean affairs.

第二十二章　治理雾霾须关注舆论的"风"

2016年年末，全国多地出现入秋以来最严重雾霾，全国40个城市发布相应级别重污染天气预警；2016年12月31日，北京启动新一轮空气重污染橙色预警，直至2017年1月7日才解除，创下北京空气重污染预警的最长纪录。随着雾霾的持续爆发，舆论场异常热闹。雾霾已不是简单的环境保护问题，而是社会和政治问题。"晨起第一件事就是看PM2.5的浓度""防霾口罩与空气净化器卖到脱销"，空气污染已成为生活常态，直接降低居民的"健康生活安全感"，如果不进行妥善处理，将严重损害政府权威和公信力。

政府监管、环境监测、媒体监督、全民关注形成了全社会积极应对雾霾的格局。然而，雾霾来临之日，便是谣言四起之时，空气指数与舆情面临双重"爆表"。各种以偏概全、错漏百出、哗众取宠的言论不断出现，甚至病毒式传播，形成非理性、非科学、非健康的舆论导向。"汽车尾气比空气干净10倍""雾霾存在硫酸铵时才会发布红色预警""风电站防护林阻挡大风导致雾霾""雾霾不散是因为核污染"等"雷人"言论，以骗取点击量为目的，造成大众心理恐慌，混淆视听，严重动摇了全社会治霾的信心。因此，治理雾霾的同时必须积极应对谣言，以舆论清风消除心中之"霾"。

谣言的主要类型

雾霾天气频发，传言趁势而起。舆论场中，真假混杂，煞有介事的"雾霾真相"和极富煽动性的帖子铺天盖地。2016年12月30日，环保部门联合释疑"2016年度十大雾霾传言"，普通民众才得以拨云见日看见真相。关于雾霾的谣言包括以下几种类型。

信口雌黄型。《别拿雾霾开玩笑了，它是一级致癌物》打着钟南山院士的旗号，称北京肺癌发病率远高于全国平均水平，呈现年轻化趋势，80个PM2.5微粒可以堵死一个肺泡、雾霾会让鲜肺6天变黑肺、吸一天雾霾就可能导致偏瘫、雾霾会导致不孕不育、雾霾让人折寿 5 年半。事实却是，钟南山院士发布声明，称"雾霾致癌的资料有不少是断章取义，夸大其词或肆意篡改"，澄清上述文章存在概念错误，非本人所写，并就引用相关数据致歉。相关数据显示，2003～2012 年，北京肺癌年平均增长率为 1.2%，2011 年肺癌年龄标准化发病率为 23.53/10 万，全国可比的最新肺癌标化发病率为 25.34/10 万，可见北京市肺癌发病率略低于全国平均水平，且发病中位年龄从 2002 年的 69 岁增长到 71 岁，年轻化趋势并不明显。

以偏概全哗众取宠型。稍有事实依据的谣言多为以偏概全哗众取宠型，网传视频用 4000 流明灯光微距镜头显示出北京雾霾，视频中一些细小颗粒四处飘散，令人不寒而栗。事实上，形成雾霾的雾滴、细颗粒物都是肉眼无法看到的，需要借助显微镜仪器，视频中所见只是灰尘而已。还有谣言称雾霾频发，北京空气质量逐步恶化。但环保部监测数据显示，截至 2016 年 12 月 27 日，2016 年北京市 PM2.5 平均浓度为 72 微克/立方米，同比下降 10.0%。联合国环境规划署 2015 年发布的评估报告显示：1998～2013 年，北京二氧化硫、二氧化氮和可吸入颗粒物的年均浓度分别显著下降了 78%、24% 和 43%，15 年间空气质量持续改善。

跨界发言偷梁换柱型。瑞典哥德堡大学抗生素耐药性研究中心 4 位学者在研究中提到，"从北京雾霾中检测出抗生素耐药性基因"，引发关注。随后国内部分微信公众号发表题为《北京雾霾中发现有耐药菌，"人类最后的抗生素"对它束手无策》《北京雾霾中含有耐药菌 60 余种将导致药物失去作用》等文章，造成了舆论的恐慌。事实上，细菌的耐药性和致病性是完全不同的概念，耐药性的增加并不意味着致病性的增强。国内外多位专家表示，细菌耐药性的获得是由进化选择和抗生素等诱导选择引起的，雾霾不产生耐药基因，雾霾与耐药菌并无必然的因果联系。

谣言因何而起

关于雾霾的谣言或断章取义，或夸夸其谈，或肆意篡改，或夸张滑稽，极容易被广泛轻信与传播,对雾霾的治理有百害而无一利。产生谣言的原因是多方面的,信息公开程度、雾霾治理效果等都会影响舆论的走向。

一是信息公开不充分是谣言产生的根源。谣言的传播根源在于政府对相关重要信息公开不及时、发布渠道不通畅,甚至对信息进行遮掩和回避。雾霾来袭,常常引起公众的无奈、疑惑、焦虑、恐慌、愤怒等负面情绪,而娱乐化、情绪宣泄式的谣言借助微博、微信等自媒体的传播手段使得不良情绪持续发酵蔓延。真相的缺失加之社交媒体的发达,造成谣言四处传播,"线上雾霾"实难"吹"走。

二是治理效果与公众感受不符给了谣言传播可乘之机。虽然污染物浓度一直处在下降趋势,但仍与公众的直观感受不符,治霾的工作和成效达不到社会的预期。例如,虽然数据显示北京的空气质量水平趋向好转,但在供暖季或遭遇极端气象条件时,雾霾仍较为严重,对居民工作生活造成很大的影响。这种雾霾治理效果与公众期待存在的差距让公众对治霾充满希冀又倍感无奈,使得舆论环境更为混乱。

三是认识不深入、科普不到位进一步催生了舆情。雾霾形成机理极其复杂,尤其在机动车是不是主要的污染源的问题上,民间和官方之间长期存在争议。随着有车族的逐渐增多、限号带来的诸多不便,民间对于"机动车排放是北京主要污染来源"的官方说法越来越倾向于反对的态度,认为污染企业、燃煤、规划不当才是污染的罪魁祸首,限制机动车对治污作用并不大。官方和民间对于雾霾成因的理解和认知不同,也为全社会就雾霾治理达成共识、形成合力带来阻碍。目前由于缺乏广覆盖、高精度的雾霾监测体系,导致发生预测预警"临时抱佛脚",结果不准确、时间不及时,甚至政府部门发布信息口径不一致,导致公众无所适从。在雾霾问题上,科学观点在舆论场中往往处于弱势,一些研究成果极易被媒体断章取义和普通民众误读,引发恐慌和焦虑。一些学者和知名人士关于雾霾的公开言论和观点也往往推波助澜,有些言论不但无益于凝聚共识、联动治霾,反而加剧了舆论的撕裂。

四是应对措施不力引发次生舆情。"临时性的治理措施",包括区域工厂停产、车辆单双号限行、停止工地作业等"短期手段",被认为是"有成效、无长效"。"碎片化的解读"使得政府的决心演化为对政府"提头来见"的片面化拷问。进而形成政府"说了不算，诚信堪忧"的负面形象，损害政府的公信力，甚至是执政能力。而学校停课、高速公路封闭、航班延误等措施，则进一步加剧了次生舆情的发生。

积极应对重霾之下的舆情

雾霾已成为国人的"心肺大患"，必须谨防雾霾来临之时舆情"爆表"。治霾之路漫长，面对大面积、持续性的雾霾天气，不能只是在吐槽、抱怨中"等风来"，政府和公众都应该有反思、有行动。

第一，增强信息供给水平，让真相走在谣言的前面。重霾之下，无人可以置身事外。政府应开诚布公，主动承担信息发布、答疑解惑的责任，最大限度地满足公众知情权、参与权、表达权和监督权，用科学观点引导公众舆论，政府要"主动说""抢着说""说得对""说得好"，提升公众"安全感"。尤其是在交通出行、上班上学、健康防护等民生方面，要细化信息内容，使公众的生活有据可循。要发挥官方网站、自媒体等"谣言粉碎机"的功能，对有关雾霾的不实言论和"伪科学"观点予以坚决、快速地回击，驱散舆论场中的"雾霾"，让公众能够以更成熟的心态面对雾霾。

第二，提升科学研究和科学普及能力，让民众正确看待雾霾。"冰冻三尺非一日之寒"，雾霾需要长期治理，并非朝夕可根治。因此，加强雾霾监测的精细化、深入化和公开化，从而为科学分析提供更为精确的数据支撑就显得尤为重要。要加强科普宣传，提升信息公开化、易懂性和亲民度，用公众容易理解、喜闻乐见的方式宣传普及雾霾的成因、危害、治霾成效与措施，甚至治霾的难点。要赢得更多的民意支持和认同。强调源头治理、解决污染物排放是根本，切勿盲目夸大和推崇"治霾神器"，以免激发公众的反感和抵触情绪。

第三，增强政府对舆情危机应对能力，争取科学的舆论话语权。提高预警水平，雾霾预报要准确，具有公信力。在发布高等级预警时，应充分统一口径、细化内容，让公众感受到应急机制清晰有序、层次分明。政府应及时回应社会上有关雾霾危害、影响、治理进度、治理理念、治理决心等热点关切，既要以理服人，又要懂得换位思考，多站在公众角度考虑问题解释问题。这不但能帮助政府取得舆论话语权，树立阳光负责的形象，也能防止网上各类谣言和炒作。回避矛盾，不愿、不敢、不会与外界正面对话，只会让负面舆情滑向难以控制的深渊。

第四，加强舆论监管，稳定情绪，齐心治霾。由于网络谣言制造成本低、传播速度快、难以揪出谣言源头，难以控制，常常陷入"先转发、再恐慌、后辟谣"的窠臼，使得公众认识更为混乱、治理信心动摇。必须加大处罚力度，提高谣言制造的成本，净化公共问题的舆论空间。政府要强化各种公众平台的事先审查义务，及时有效阻击谣言的传播，稳定公众情绪，增强公众对治霾的信心。

Chapter 22
Pay Attention to Public Opinions on Smog Control

At the end of 2016, the most serious smog since autumn struck many areas of China. More than 40 cities issued different levels of heavy pollution alerts. On December 31, 2016, Beijing launched a new round of heavy air pollution alert, which continued until January 7, 2017, creating a record of the longest heavy air pollution alert. With the continuous outbreak of smog, public opinions are overheated. Smog is not only a simple environmental protection issue, but also a social and political issue. "The first thing in the morning is to check the concentration of PM2.5". "Pollution masks and air purifiers are sold out". Air pollution has become a normal scene in our life, directly reducing people's sense of security on healthy living. If not handled properly, it will seriously damage the authority and credibility of the Chinese Government.

Today response to smog includes government regulations, environmental monitoring, media supervision and concerns of the people. However, every time when smog hits our country, rumors will follow. All kinds of partial, misleading and grandstanding statements go viral, leading to irrational, unscientific and unhealthy public opinions. "The car exhaust is 10 times cleaner than the air". "Red alert will only be triggered when there is ammonium sulfate in the smog". "Smog is caused by wind power station shelters because they have blocked wind". "Smog is caused by nuclear pollution". These kinds of ridiculous opinions aim at gaining more internet hits, causing public panics and seriously shaking the confidence of the whole society in smog treatment. Therefore, we need to take active moves to rumors when dealing with smog so as to remove doubts of the people.

Main Types of Rumors

Rumors rise up with the outbreak of frequent smog. False and true information mixed in public opinions. Inflammatory statements and posts are stirring up public opinions. On December 30, 2016, environmental protection departments at all levels jointly refuted ten main rumors about smog in 2016 so that the general public was able to see the truth. Rumors about smog mainly have the following features.

First, rumors out of thin air. An article under the false name of academician Zhong Nanshan, *Be Serious about Smog for It Is a Primary Carcinogen*, said the incidence of lung cancer in Beijing is much higher than the national average and more young people are suffering from it. It also said 80 PM2.5 particles can block an alveolar, smog will turn lungs to black within six days and smog may lead to hemiplegia and infertility and shorten people' life by 5 years. However, academician Zhong Nanshan issued a statement that "much information indicating smog is carcinogenic is out of thin air, exaggerated or tampered randomly" to clarify that the above article has a conceptual error and is not written by himself. He also apologized for referring to relevant data. Relevant data show that from 2003 to 2012, the average annual growth rate of lung cancer incidence in Beijing was 1.2%. In 2011, the standardized incidence of lung cancer was 23.53/100,000, while it was 25.34/100,000 across the country. So the incidence of lung cancer in Beijing is slightly lower than the national average. What's more, median age of people with lung cancer rose from 69 years old in 2002 to 71 years old, refuting the idea that more young people are suffering from it.

Second, partial and grandstanding rumors. Most of rumors based on partial facts are grandstanding. A video shot with 4000 lumens of light and macro lens shows smog in Beijing. In the frightening video, small particles are floating around. In fact, droplets and fine particles in the smog are invisible to naked eyes and people can only see them by using microscopic instruments, so the particles in the video are only dusts. Some rumors said that smog frequently happened in Beijing and the air quality is declining. However, monitoring data of the Ministry of Environmental Protection show that as of December 27, 2016, the average concentration of PM2.5 in Beijing was 72-micrograms/cubic meter in 2016, down 10.0% year on year. The assessment report released by the United Nations Environment Program (UNEP) in 2015 showed that the annual average concentrations of sulfur dioxide, nitrogen dioxide and respirable particulates matters in Beijing significantly reduced by 78%, 24% and 43% respectively from 1998 to 2013, during when the air quality was improved.

Third, rumors that tamper the source information. Four scholars at the Centre for Antibiotic Resistance Research at University of Gothenburg mentioned in a study that they have "detected antibiotic resistance gene from the smog sample of Beijing", raising concerns among people. Some WeChat accounts issued articles such as *Resistant Bacteria Was Found in Beijing Smog and The Last Human Antibiotics Is Useless to It, More Than 60 Kinds of Resistant Bacteria in Beijing Smog Make Drugs Lose Efficacy*, causing panic among the public. In fact, bacteria's resistance and pathogenicity are completely different things and higher drug resistance does not lead to increased pathogenicity. A number of experts at home and abroad said bacteria obtained resistance by adaptive evolution and antibiotics induced selection, and smog does not produce resistance genes and has nothing to do with drug-resistant bacteria.

Causes of Rumors

Rumors about smog are out of context, exaggerating, tempered randomly or ridiculous. They will go viral and people will easily get deceived, so they will do no good to smog treatment. The causes for the rumors are multifaceted and the degree of information disclosure and the effect of smog treatment will both affect the trend of public opinions.

First, inadequate disclosure of information is the root cause of rumors. The root causes is that the government's disclosure of relevant important information is not timely, the distribution channels are not smooth, and the government even cover and avoid some information. Smog often makes people helpless, confused, anxious, panic and angry. Rumors aiming at entertaining and unleashing fury will spread those negative emotions around through self-made media such as Weibo and WeChat. The lack of truth and the development of social media resulted in widespread rumors that make it hard to remove people's doubts.

Second, rumors can take advantage of unsatisfactory smog treatment and public feelings. Although the concentration of pollutants has been in a downward trend, it is still inconsistent with the public intuitive feelings and the result of smog treatment didn't meet social expectations. For example, although the data show that Beijing's air quality tends to improve, in the heating season or extreme weather conditions, smog is still very serious, exerting a great impact on people's life and work. Smog treatment gave people hopes and let them down, making public opinions more chaotic.

Third, lack of understanding and insufficient education further gives opportunities to rumors. Smog formation mechanism is extremely complex. In particular, civil and official opinions have long diverged in the issue of whether motor vehicle is the main source of pollution. With the increase in the number of car owners and restriction of cars on the roads, more people tend to oppose the idea that motor vehicle emission is the main source of pollution in Beijing. They think the root cause is polluting enterprises, coal burning and improper planning and restriction of cars on the roads is of not so much use. Divided understandings and opinions on the causes of smog between the official and civil side become an obstacle for whole society to reach a consensus and work together on the smog treatment. At present lack of wide coverage and highly precise smog monitoring system results in untimely and imprecise early warning. Sometimes, even information released by different government agencies is inconsistent, further making people confused. The scientific point of view concerning smog is often ignored and some research results can easily be misinterpreted by ordinary people, causing panics and anxiety. Public remarks and views on smog made by some scholars and celebrities often fuel public concerns, doing no good to achievement of consensus and joint actions and even increasing controversies.

Fourth, inefficient measures will let spreading of rumors go unchecked. Temporary measures, including the regional plant shutdown, restrictions of cars on the roads, halting of site operations, are considered ineffective and unsustainable. Due to partial interpretation, the government's determination was misunderstood as "apologizing for the unsatisfactory effect by death". People then hold negative view on the government and cast doubts on the government's credibility and even its ability to govern. Some measures such as suspending classes, blocking highways and delaying flights, further exacerbate the spreading of rumors.

Responding Actively to Public Opinions on Heavy Smog

Smog has raised great concerns among people about their health. We must be cautious about overheating of public opinions. It takes time to treat smog, so confronting with extensive and long-lasting smog, the government and people need to reflect on themselves and take actions instead of just complaining and waiting for the wind.

First, we will provide more information to let people know the truth before the spread of rumors. No one can stand aside in front of heavy smog. The government should be open and honest to the public, take the initiative to disseminate information and respond to people's doubts so as to meet people's rights to know, participate in, express and monitor. The government should guide public opinions with scientific points of view. The government needs to release true information on time and even ahead of time to enhance people's sense of security. Detailed information concerning transportation, work, school and health should be released to get people informed. We will make use of official websites and self-created media to crush rumors, swiftly fight back false statements and points of view and remove doubts in public opinions, so that the public can hold a more mature attitude to smog.

Second, we will enhance scientific research and capability of public education so that people can hold rational views on smog. As the old saying goes, Rome is not build in one day. Smog cannot be treated overnight. Therefore, it is particularly important to enhance the refinement, deepening and transparency of smog monitoring and provide more accurate data for scientific analysis. We will promote public education, information transparency and readability, and popular publicity methods to make people understand the causes and danger of smog, the results of smog treatment and even difficulties in dealing with smog. We need to win more public support and recognition. We will focus on removing the source of pollution and reduce pollutant emissions. We shouldn't blindly promote the so called "all-powerful smog removal machines" lest we may stimulate resentment and resistance from the public.

Third, the government's ability to deal with public opinion crisis will be improved to increase its voice in public opinions. The quality of warning will be raised to release more accurate and credible smog forecasts. High-level warning should offer consistent and detailed information so that the public can see a clear, orderly and structured emergency mechanism. The government should promptly respond to social concerns about the danger and influence of smog, treatment progress and philosophy, and the government's determination. The government should not only convince people by reasoning, but also think in people's shoes and consider and explain the issues from people's perspective of view. This will help the government increase its voice in pubic opinions, build a credible image and prevent all kinds of online rumors and speculation. To avoid contradictions, be unwilling to and dare not to face a positive dialogue with the outside world will only let the negative public opinions go out of control.

Fourth, we will strengthen regulations on public opinions and calm down and unify people to cope with smog. Low cost, fast spreading and difficulties in finding the source make it harder to control rumors. We are often trapped in a vicious circle of "forwarding, panicking over and refuting the rumors", making public awareness more chaotic and shaking people's confidence in smog treatment. We must toughen punishment, increase cost of rumor creating and clean up public opinions on public issues. The government should strengthen beforehand examination of various public platforms, promptly and effectively block the spread of rumors, stabilize public sentiment and enhance public confidence in smog treatment.

第二十三章　提升水治理能力　迎来河清海晏

水是生命之源，与人类文明息息相关。文明的生存与发展必然要求兴水利、除水害，正如古人指出的那样——"水利兴，而后天下可平"。近年来，我国的水治理工作在取得一定成效的同时，也面临一系列亟待解决的问题。正确认识和梳理这些问题，对于有效减少旱涝灾害、合理开发利用水资源、保护水环境，进而对实现海晏河清有重大意义。

治理成效初显　存量问题突出

防汛抗旱减灾成效总体显著，局部存在薄弱环节。通过大力修建防洪工程、抗旱水源工程和灌溉工程，防汛抗旱减灾能力提高。2015 年全国水库总库容达 9323 亿立方米，保护耕地 7 亿多亩，保护人口 5.98 亿人。七大江河初步形成了以水库、堤防、蓄滞洪区为主体的拦、排、滞、分相结合的防洪工程体系。在抗旱方面，开源与节流并重。近年来，在节流方面成效突出，全国建成 100 个节水型社会试点，农业推广高效节水灌溉，工业实施高耗水行业节水技术升级改造，生活服务业加快供水管网改造，推广节水器具。国内防汛抗旱的薄弱环节主要包括水利设施不够完备的中小河流、城市内涝等。

水资源保障能力稳步提高，部分地区仍面临水资源短缺。水资源保障能力提高体现在三个方面。一是供水能力大幅提升。如南水北调工程纵贯长江、淮河、黄河、海河四大流域，总调水规模 448 亿立方米，为数以亿计的北方人民带来清洁、稳定的饮用水源。二是新型水资源开发利用技术助力水资源保护。这些新型技术包括城市雨水资源收集利用、流域洪水资源安全利用、污水资源回收利用以及海水淡化等。三是农村饮用水安全稳步提高，农村集中式供水人口比例提高到 58% 以上，先后解决了超过 6 亿农村人口的饮水安全问题。然而，华北、西北等地区缺水严重，同时全国 400 多座城市缺水，由此带来地下水超采严重，全国每年超采地下水约160亿立方米。

水污染控制能力逐渐提升，既有污染形势严峻。一方面，我国已成为全世界污水处理能力最大的国家之一。截至2015年年末，我国的城镇污水日处理能力由2010年的1.21亿吨增加到1.82亿吨。另一方面，我国高度重视水体水质问题。2015年国务院印发《水污染防治行动计划》，对七大重点流域水质优良（达到或优于Ⅲ类）比例提出要求。2015年，全国972个地表水国控断面（点位）的监测结果表明，水质优良比例达到64.5%，比2014年提高了1.4个百分点。但是，国内仍有40%左右的河湖存在严重污染，农村面源污染严重，地下水污染也不容乐观，水质较差和极差的地下水占到57.3%。

制度体系与治理能力有待加强

行政管理机构权责不清，部分领域"九龙治水"。目前，在水治理的多个领域存在管理部门权责不清的问题。总体来看，水利部门是水行政主管部门，然而实际涉水事务中并非水利部门一家。如水功能区划拟定方面，《中华人民共和国水法》（以下简称《水法》）规定"水行政主管部门会同环保部门拟定"，然而实际操作中环保与水利部门职责并不清晰，一个后果是我国50%以上的重要饮用水水源保护区没有划定；在水环境监测与水质监测、排污总量控制目标和水域限制排污总量限制方面，上述两个部门也存在职责不清晰、衔接不通畅的问题。在城市给排水和防洪排涝方面，水利部门与住建部门存在职能交叉，须加强协调。在湿地管理方面，水利与林业部门对湖泊湿地管理范畴存在界定不清问题。

流域综合管理能力薄弱，管理协调机制缺位。一方面，流域立法落后。专门针对流域管理的法规只有《太湖流域管理条例》《长江河道采砂管理条例》《黄河水量调度条例》《淮河流域水污染防治暂行条例》等，其他法律法规对流域管理职能的规定过于原则化，对流域综合管理支撑不足。另一方面，流域管理协商机制不健全。目前长江、黄河等河流或湖泊已有水利部派出的流域机构，然而这些机构缺乏财政权、规划权与监管执法权，难以谋划全流域水资源的开发、利用、保护和水污染的治理。同时流域内地方政府间缺乏必要的协商机制，也会带来一系列消极后果。

治理实践与科技结合不足，存在一定的认知偏差。一是水资源保护重地上、轻地下。目前，我国水污染的范围由流域污染向湖泊、地表水、地下水蔓延，地下水超采情况严重，我国北方地区因地下水超采形成了区域地下水降落漏斗。二是污水处理重城市、轻乡村。目前农村地区污水处理率仅为10%左右，大量未经处理的农村污水直接排入河道水系，成为主要的水污染源。三是水污染防治重污水、轻污泥。住建部数据显示，当前污水处理过程中，约56%的城市污泥未做到无害化处理。

完善治理机制　实现科学治水

治水兴水核心在于推进水治理体系和治理能力现代化。在贯彻落实"节水优先、空间均衡、系统治理、两手发力"的治水思路的基础上，结合我国水治理体制的实际，可以从如下几个方面完善治理机制。

完善水治理制度体系，强化治理法治保障。首先，要构建系统、完善、科学规范的水灾害防治、水资源保障、水污染防治的法律体系。适时启动《水法》和《中华人民共和国防洪法》(以下简称《防洪法》)等法律的修订工作；完善农田水利、饮用水安全保障等方面的水法规；完善节约用水、地下水管理、水权交易等方面的水法规；完善河湖管理、河道采砂、水利工程管理与保护等方面的水法规。其次，明晰涉水部门的职能分工，提高行政效能。借鉴发达国家的经验，适应生态文明体制改革要求，分别明确水利、环保、住建、海洋等部门在水资源、水环境、城镇涉水事务、海洋保护等方面的职能分工，理顺部门之间的职责关系。最后，做好流域综合立法和有关单项立法工作，明确流域机构的法律和行政地位，强化流域管理机构在流域综合规划编制、防洪安全管理、水资源统一调度、"三条红线"控制指标考核评估等方面的职能；同时明确流域机构与行政区域的职能划分，流域机构主要负责国家水安全战略和重大水利规划、政策的贯彻落实和跨流域、跨区域、跨国界河流湖泊以及事关全局的涉水事务，其余涉水事务由行政区域负责。

　　发挥市场机制作用，引入各方协同治理。首先，切实转变涉水行政部门的政府职能，加大政府购买涉水公共服务力度。如推动社会资本参与城镇供水、污水处理项目建设运营，缓解地方政府资金不足难以支撑水治理资金需求的困境。其次，建立健全科学合理的水价机制，发挥价格杠杆在水资源配置中的作用。积极探索多种形式的水权交易流转方式，因地制宜探索地区间、流域间水权交易流转方式。最后，鼓励企业、公众等主体参与水治理。建立水治理议事协调机制，保障一定比例的利益相关者和公众代表参与水治理决策，加强宣传教育等手段确保公众和利益相关者的有效参与。提高企业和社会团体参与水治理的法律地位，明确在涉水事务中的职责、权利和义务，同时加强政策引导，充分发挥龙头企业的规模效应，推动水务行业市场化整合，优化行业资源配置。

　　鼓励涉水事务科技创新，依靠科技治水兴水。一是加强涉水事务基础科学研究，使治水符合自然生态规律、经济发展规律和社会发展规律，防止出现以往水资源保护重地上、轻地下等问题。二是鼓励涉水技术创新。在防汛抗旱方面，建设治水工程要学习都江堰等治水技术创新的经验，强化工程技术的前瞻性，使得工程项目不落伍；在水资源保护方面，促进和完善工业循环用水系统，改革生产工艺和用水工艺，建立和完善城市再生水利用技术体系；在水环境保护方面，推进水文监测、水质监测、水量监测等领域新技术的利用，不同监测体系的有效整合，促进监测信息的及时发布、传输和运用，实现智慧治水、数字治水、科学治水。

Chapter 23
Enhance Water Governance Capabilities

Water is the source of life and is closely linked to human civilization. The survival and development of civilization is bound to require building water projects and eliminating water disasters. As the ancient people pointed out,"only if the water projects are built, can we ensure social peace". In recent years, China's water governance work has also been confronted with a series of problems to be settled urgently while achieving certain results. It is of great significance to have a correct understanding and combing of these problems for effectively reducing drought and flood disasters, making reasonable development and utilization of water resources and protecting the water environment, thus to achieve social peace.

Great Success Has Been Achieved, but Problems Remain

Generally speaking, flood control and drought relief has achieved remarkable results, but there are local weak links. Through great efforts to build flood control projects, drought and water conservancy projects and irrigation projects, flood control and drought reduction capacity has been improved. In 2015, the total storage of the national reservoirs reached 932.3 billion cubic meters, more than 700 million mu of arable land were protected, as well as a population of 598 million. Seven rivers have initially formed a flood control engineering system with reservoirs, dikes and flood storage and detention areas as the focus, integrating the functions of water retaining, drainage, retention and diversion. In terms of drought control, equal importance is attached to opening up the source and regulating the flow. In recent years, outstanding achievements have been made in flow regulation. The country has established 100 national water-saving social pilot projects to promote efficient water-saving irrigation in agriculture, upgrade and the water-saving technology in the high water consumption industry and speed up water supply pipe network reconstruction and promote water-saving appliances in life service industry. In addition, the weak links of flood control and drought in China mainly include small and medium rivers with insufficient water conservancy facilities and waterlogging in cities, etc.

Water resources guarantee capacity has steadily improved, but some areas still face water shortages. The improvement in water resources guarantee capacity is mainly reflected in three aspects. First, water supply capacity is significantly increased. For example, the South-North Water Transfer Project runs through four basins of the Yangtze River, Huaihe River, Yellow River and Haihe River, with a total water transfer capacity of 44.8 billion cubic meters, bringing clean and stable drinking water for hundreds of millions of people in the north. Second, the new technologies to develop and utilize water resources boost protection of water resources. These new technologies include collection and utilization of rainwater resources, safe utilization of flood resources in the basins, recycling of sewage resources and desalination, etc. Third, safety of the drinking water in rural areas is steadily improved and the proportion of the population with centralized water supply in rural areas has been increased above 58%, which has successively solved the water safety issues for more than 600 million people in rural areas. However, North China, Northwest and other areas are confronted with serious water shortage, while more than 400 cities nationwide lack water, resulting in serious groundwater overdraft of about 16 billion cubic meters of underground water in the country every year.

Water pollution control capacity has been gradually increased, but the pollution situation remains serious. On the one hand, China has become one of those countries in the world with the largest sewage treatment capacity. As of the end of 2015, China's daily urban sewage treatment capacity increased from 121 million tons in 2010 to 182 million tons. On the other hand, China attaches great importance to water quality problems. In 2005, the State Council printed and issued the *Action Plan for Prevention and Control of Water Pollution* to made requests for the proportion of quality water (reaching or better than Grade III) in seven key river basins. In 2015, monitoring results of 972 surface water state-controlled sections (points) in the country showed that the proportion of quality water reached 64.5%, increased by 1.4% as compared with 2014. However, still about 40% of rivers are facing serious pollution, non-point source pollution is serious in rural areas and groundwater pollution is not optimistic, and the proportion of underground water with poor and extremely poor water reaches up to 57.3%.

Improving the Water Governance System and Capabilities

The fact is that the power and responsibility of the administrative departments are unclear and some fields are put under the administration of multiple departments. At present, the problem of unclear division of power and responsibility can be found in the administrative departments in many fields of water treatment. In general, the water conservancy department is the competent administrative department of water affairs, but actually it is not the only department involved in water affairs. For example, with regard to formulation of water functional zoning, *Water Law of the People's Republic of China* stipulates that "it shall be formulated by the competent administrative department of water affairs jointly with the environmental protection department" However, the responsibilities of the environmental protection department and water conservancy department are not clear in the actual practice, as a result of which more than half of the important drinking water conservation districts in the country are not designated. In terms of water environment monitoring and water quality monitoring, control object of total pollutant amount and cap of limited total pollutant amount in water area, problems can still be found in the above two departments, such as unclear responsibilities and unsmooth coordination. In terms of urban water supply and drainage and flood prevention and drainage, the Water Conservancy Department and the Housing Construction Department also have overlapping functions, which needs to strengthen coordination. In terms of wetland management, the water and forestry sectors cannot clearly define the scope of management of lakes and wetlands.

Integrated watershed management capacity is weak and management coordination mechanism is vacant. On the one hand, the watershed legislation is behindhand. Specific regulations on watershed management include only *Regulations on Taihu Watershed Management, Regulations on Management of*

Sand Excavation in Yangtze River Channel, *Regulations on Water Scheduling of the Yellow River*, *Interim Regulations on Prevention and Control of Water Pollution in the Huai River Basin* and so on, because other laws and regulations seem to be too principle for the regulations on watershed management functions and lack support for integrated watershed management. On the other hand, the watershed management negotiation mechanism is not sound. At present, the Yangtze River, the Yellow River and other rivers or lakes have the basin institutions assigned by the Ministry of Water Resources, but these institutions lack the financial power, planning power and regulatory enforcement power, so it is difficult for them to plan the development, utilization and protection of whole basin water resources and water pollution control. At the same time, the lack of necessary negotiation mechanism among local governments in the basin also brings a series of negative consequences.

The combination of governance practice and science and technology is insufficient, leaving certain cognitive biases. First is to attach importance to ground protection but look down on underground protection of water resources. At present, the scope of water pollution in China has spread from watershed to lakes, surface water and underground water. Given the serious situation of groundwater overdraft, regional groundwater depression cone has been formed in North China due to groundwater overdraft. Second is to attach importance to urban areas but look down on rural areas in terms of sewage treatment. At present, the sewage treatment rate in rural areas is only about 10%, and a large amount of untreated rural sewage is directly discharged into the river water system, becoming a major source of water pollution. Third is to attach important to sewage but look down on sludge in terms of prevention and control of water pollution. Data provided by the Ministry of Housing and Construction show that about 56% of urban sludge is not provided with harmless treatment in the current sewage treatment process.

Improving the Governance Mechanism to Achieve Scientific Flood Control

The key of water control and building water conservancy projects is to advance the modernization of water governance system and governance capabilities. Based on implementing the water control ideas of "water conservation first, spatial equilibrium, system governance and exerting best effort", we can improve the governance mechanism from the following aspects in combination with the actual situation of the water governance system in China.

Actions shall be taken to improve the water governance system and strengthen legal guarantee for governance. Firstly, we shall establish a systematic, complete and scientific and standardized legal system for prevention and control of water disasters, water resources guarantee and prevention and control of water pollution. We shall initiate revision of *Water Law*, *Flood Control Act* and other laws in due time; improve the water regulations in terms of irrigation and water conservancy, safety guarantee of drinking water and so on; improve the water regulations in terms of water conservation, groundwater management, exchange of water rights and so on; improve the water regulations in terms of river and lake management, sand mining in river channel, water project management and protection and so on. Secondly, we shall make clear the division of functions of the water-related departments to improve administrative efficiency. We shall learn from the experience of developed countries to adapt to the requirements of ecological civilization system reform, and respectively clarify the functional division of water conservation department, environmental protection department, housing and construction department, marine department and other departments in terms of water resources, water environment, urban water-related affairs, ocean protection and so on to straighten up the responsibility relationship between those departments. Finally, we shall do a good job in integrated watershed legislation and relevant single legislation, make clear the legal and administration positions of watershed institutions and enhance the functions of the watershed management institutions in integration watershed planning preparation, flood control safety management, unified scheduling of water resources, assessment and evaluation of control indicators of "Three Red Lines". At the same time, we shall clarify the functional division of watershed institutions and administrative regions, of which the watershed institutions are mainly responsible for implementation of national water security strategy and major water conservancy planning and policies and cross-basin, cross-region and cross-border lakes and rivers as well as water-related affairs in relation to the overall situation, while the administrative regions are responsible for other water-related affairs.

Measures shall be taken to display the role of market mechanism and engage the parties concerned for co-governance. Firstly, we shall effectively change the governmental functions of water-related administrative departments and devote greater efforts for the government to purchase water-related public services. For example, we should take measures to promote the participation of private capital in construction and operation of urban water supply and sewage treatment projects to relieve the dilemma that the insufficient local government capital makes it difficult to satisfy the fund demand for water governance. Secondly, we should establish and improve a scientific and rational water pricing mechanism to give play to the role of price leverage in allocation of water resources. In addition, we should also actively explore a variety of forms for the transfer mode of water rights and explore the transfer mode of water rights between different regions and basins according to the local circumstances. Finally, we should encourage enterprises, the public and other players to participate in water governance. We shall establish a deliberation and coordination mechanism for water governance, ensure a certain proportion of stakeholders and public representatives to participate in water governance decision-making and strengthen publicity of education and other means to ensure the effective participation of the public and stakeholders. We shall also improve the legal status of enterprises and social groups participating in water governance, clarify their duties, rights and obligations in water-related affairs, and strengthen policy guidance, give full play to the scale effect of leading enterprises and promote market integration of the water industry to optimize allocation of industry resources.

Actions shall be taken to encourage technical innovation in water-related affairs and control water and build water conservancy projects by virtue of science and technology. Firstly, we shall strengthen the basic scientific research on water-related affairs, so that water control will be in line with the laws of natural ecology, economic development and social development, thus to prevent the previous problems such as attaching importance to ground protection while looking down on underground protection of water resources and so on. Secondly, we shall encourage water-related technical innovation. In the aspect of flood control and drought relief, we shall learn from the experience of technical innovation for water control in Dujiangyan and other places if we want to build water conservancy projects, and enhance the perspectiveness of engineering technology to enable the engineering project not to fall behind. In the aspect of water resources protection, we shall promote and improve the industrial circulating water system, innovate the protection technology and water-using technology and establish and improve the urban recycling water utilization technology system. In the aspect of water environmental protection, we shall promote the application of new technologies in hydrological monitoring, water quality monitoring, water quantity monitoring and other fields, effectively integrate different monitoring systems and promote timely release, transmission and use of monitoring information to achieve water control in an intelligent, digital and scientific manner.

第二十四章　防治土壤污染　守护沃土良田

土壤本身的特性决定了其污染易、修复难、投入大、见效慢。可以预见，土壤污染防控必然是我国政府向污染"宣战"的重要一役，既是持久战又是攻坚战，必须重视建立长效机制，积极削减存量、严格管控增量，强化标本兼治，加快推进土壤污染持续修复治理工作。

"日月丽乎天，百谷草木丽乎土"，土壤承载着人类与万物的生存繁衍。然而长期以来，经济社会快速发展，土壤污染却如影随形，虽经政府着力治理，污染形势仍然严峻。除了传统的农用地土壤污染问题以外，近年来产业转型升级过程中一些化工、冶金等传统污染企业的搬迁或关闭，又使工业污染场地问题凸显出来。土壤污染问题广泛关涉工业农业生产，影响国计民生乃至文明延续，土壤修复迫在眉睫、势在必行。

土壤污染严重　修复进度缓慢

我国土壤环境状况总体不容乐观，耕地和工矿用地污染较为严重，污染物以重金属等无机物为主。根据《全国土壤污染状况调查公报》，从总体格局上看，全国土壤总超标率达16.1%，其中轻微、轻度、中度和重度污染点位比例为11.2%、2.3%、1.5%和1.1%；从土地利用类型看，耕地、林地、草地土壤点位超标率分别为19.4%、10.0%、10.4%，重污染企业及周边土壤点位超标率达36.3%，固体废物集中处理处置场地土壤点位超标率达21.3%，采矿区土壤点位超标率达33.4%；从污染物构成看，镉、镍、砷分别以7.0%、4.8%、2.7%的点位超标率成为最主要的污染物，滴滴涕、多环芳烃等污染物点位超标率也相对较高。随着资源能源消耗和人为活动强度的持续增加，土壤环境保护面临的压力有增无减。

在土壤环境状况不容乐观的条件下，修复工作任重道远。当前，土壤修复已得到各级政府的重视，在国务院颁布《土壤污染防治行动计划》（又称《土十条》）的基础上，福建、安徽等20个省区市已公布地方版《土十条》，在分解细化国家标准之外，突出地方土壤环境特色：如黑龙江、吉林两省将黑土地列为治理重点之一。然而，当前我国土壤修复能力仍然受到技术和资金的制约。技术方面，与大气污染和水污染治理不同，土壤修复耗时长、耗力大、处理过程更繁复，加之土壤类型多样，受体（生物、水体、空气、人体等）类型众多，故土壤修复技术也比较复杂。目前，我国不少土壤修复技术仍停留在实验室水平，尚没有行之有效的技术体系，关键设备仍须进口。资金方面，当前土壤修复资金仰赖政府部门和土地开发商，而排污企业等造成土壤污染的主体由于产权不明、规模有限等原因难以提供足够的资金，"谁污染，谁治理"的原则贯彻步履维艰、难以推进，致使我国土壤修复能力难以快速提高，增量污染问题依然严峻，"毒地"现象屡见不鲜。

制度短板制约土壤修复

"看得见的污染，看不见的制度。"当前土壤污染防治除技术性困难之外，还存在一些制度空白，这些空白成为提升土壤污染防治工作的瓶颈。弥补制度缺失，才能为治理开门引路、保驾护航。

一是污染底数不清，现有数据较为粗糙。土壤污染的"家底"是所有防治工作的前提和基础。我国尚未建立起按照污染程度进行分类的农用地清单和污染地块清单，监管依据不足。尽管《全国土壤污染状况调查公报》公布了一些土壤污染检测数据，但是调查点位设置较为稀疏，数据难以准确代表土壤污染的分布及状况，部分地区甚至缺乏监测点位和人员配置。因此，亟须以农用地和重点行业的企业用地为重心开展土壤污染状况的详查。

二是制度体系尚待完善，法治保障仍须加强。当前，我国的土壤污染法律体系还未建立起来，相关规定散见于其他环境污染法律法规中。法律体系可以预防土壤污染违法行为，然而，目前法律缺位致使法律责任、主体等认定存难，加剧了有关土壤污染违法案件的处理难度；同时，法律还可以就治污手段做出规定，目前我国较少采用经济手段，如引导市场行为的鼓励政策。美国在土壤污染的管理方面，以《环境应对、赔偿和责任综合法》为基础建立了超级基金场地管理制度，包括责任追究、修复响应和污染赔偿，也就是"谁污染，谁治理"及"谁污染，谁担责"，有力地震慑了污染者，反过来起到了污染防治的效果。此外，由于监管未实现完全的有法可依，加上各地土壤环境监测、监督执法、风险预警体系建设严重滞后，难以对辖区内土壤环境实施有效监控。

三是标准规范精细度低，技术体系亟待完善。现行的土壤环境标准不健全，使得目前土壤环境污染评价与分区都缺乏适当的依据。目前执行的《土壤环境质量标准》（1995）粗泛且针对性弱，虽然2014年环保部发布了建设用地土壤环境调查、监测、评估、修复系列标准作为补充，但总体上看还存在诸多问题。一是土壤标准制定过分要求统一，未充分考虑我国土壤类型复杂的现实情况，没有实现因地制宜；二是污染物项目较少，尤其是有机污染物种类偏少；三是重金属污染物指标注重总量控制，但土壤污染主要是由有效态构成，总量不能准确反映土壤受到的实际影响；四是现行标准的制定未充分考虑土壤对生物受体的毒性效应。

土壤修复需要政策与市场的双重助力

土壤本身的特性决定了其治理过程污染易、修复难、投入大、见效慢。可以预见，土壤污染防控必然是我国政府向污染"宣战"的重要一役，既是持久战又是攻坚战，必须重视建立长效机制，积极削减存量、严格管控增量，强化标本兼治，加快推进土壤污染持续修复治理工作。

第一，防治土壤污染，完善制度先行。首先，须结合《土十条》的实施，加快土壤污染防治立法。坚持"保护优先、预防为主、防治结合"的方针和"分类、分区、分目标管理"的原则，建立符合中国国情的土壤污染防治体系。在防治体系中重点要设立公平合理的责任认定原则，并根据不同的法律责任主体规定相应的法律责任,实践"谁污染,谁负责"原则。其次,尽快修订《土壤环境质量标准》等相关标准,如污染土壤治理与修复、重点区域行业重金属污染物排放限值、主要污染物分析测试方法、土壤标准样品等标准,以及风险评估、土壤污染治理与修复等技术规范,满足土壤环境监管工作的需要。再次,创新监管模式,提升监管能力。如在场地监管方面,根据美国新泽西州的经验,采用目标导向模式,推进实现系统的项目计划、动态的工作策略和实时的测量技术动态一体化互动的场地管理,以降低场地调查和修复过程中的种种不确定性。

第二,制定市场规则,刺激修复活力。土壤修复还要善于利于经济手段,重点在于制定市场规则、引导资源配置。在投融资、税收、补贴等方面,扶持跨区域、区域内环保设施统一规划、统一建设、统一运营,实现环保基础设施一体化管理。以市域、县域为单元,加快完善环境调查、分析测试、风险评估、治理与修复工程设计和施工等全过程产业链发展,打通"闭塞"环节,发挥"互联网+"在全产业链中的作用,促进土壤、水、大气污染协同治理产业。通过政府和社会资本合作(Public-Private Partnership,PPP)模式,加大政府购买服务力度,带动更多社会资本参与污染协同治理。发挥政策性和开发性金融机构引导作用,为跨区域、多介质协同治理的重大污染防治项目提供支持。

第三,呼唤可实践、符合现状需求的新技术。我国的污染土壤修复技术研发应该为解决农田土壤污染、工业场地土壤污染、矿区及周边土壤污染以及生态敏感的湿地土壤污染等问题提供技术支持,这就需要研发能适用于不同土壤类型与条件、不同土地利用方式和不同污染类型与程度的土壤修复技术。针对受重金属和有机物污染的农业土壤或湿地土壤,需要着力发展更安全、低成本、环境友好、能大面积应用的生物修复技术和物化稳定技术,实现边修复边生产,以保障农产品安全和生态安全。针对工矿企业废弃的化工、冶炼等各类重污染场地土壤,需要着力发展具有场地针对性、能满足安全与再开发利用目标的修复工程技术,开发具有自主知识产权的成套修复技术与设备，形成系统的场地土壤修复标准和技术规范，以保障人居环境安全健康。针对各类矿区及尾矿污染土壤，现阶段需要着力研究能控制水土流失与污染物扩散的生物稳定与生态工程修复技术，以保障矿区及周边生态环境安全，并提高其生态服务价值。

Chapter 24
Prevent and Control Soil Pollution to Safeguard Fertile Soil and Farmland

Characteristics of the soil itself make it easy to be polluted and difficult to be restored, which costs large input but is slow to take effect. It is foreseeable that the prevention and control of soil pollution will inevitably become an important campaign of the Chinese government to "declare war" against pollution. It will not only be a protracted war but also a tough fight. Therefore, we must pay attention to the establishment of a long-term mechanism, actively reduce the stock, tighten the control over the increment and strengthen seeking both temporary and permanent solutions to accelerate the continuous restoration and control of soil pollution.

As the old saying goes that "the sun and the moon run in the sky and the plants grow in the soil", it shows that soil carries the survival and reproduction of human beings and all things on earth. However for a long time, with the rapid development of economy and society, the country is still haunted by soil pollution. Although the government focuses on governance, the pollution situation is still grim. In addition to the traditional soil pollution in agricultural land, some traditional polluting enterprises in chemical industry, metallurgical industry and other industries have moved or shut down in recent years in the process of industry transformation and upgrading, making the industrial pollution site prominent again. In addition, soil pollution issues are widely related to industrial and agricultural production and affect the national interest and people's livelihood and even civilization continuation, so it is extremely urgent and imperative to restore the soil.

Slow Progress in Restoring Heavily Polluted Soil

China's soil environmental conditions in general are not optimistic, arable land and industrial and mining land are heavily polluted and the pollutants are mainly the heavy metals and other inorganic substances. According to the *Bulletin on National Soil Pollution Survey*, it can be seen from the overall pattern that the total over-standard rate of soil in the country reached up to 16.1%, of which slight, mild, moderate and heavy pollution points separately took up 11.2%, 2.3%, 1.5% and 1.1%. From the perspective of land use type, the over-standard rate of cultivated land, forest land and grassland points separately took up 19.4%, 10.0% and 10.4%. The over-standard rate of heavy-pollution enterprises and surrounding soil points reached 36.3%, that in the soil point for centralized treatment and disposal for solid wastes reached 21.3% and that of the soil point in the mining area reached 33.4%. From the composition of pollutants, cadmium, nickel and arsenic became the major pollutants with the respective point over-standard rate of 7.0%, 4.8% and 2.7%. In addition, the point over-standard rate of DDT, PAH and other pollutants were also relatively high. With the resources and energy consumption and the intensity of human activities continue to increase, soil environmental protection also faces the increasing pressure.

Under the circumstance that the soil environment conditions are not optimistic, there is a long way to go for soil restoration. At present, soil restoration has received attention from the governments at all levels. Based on the *10-Chapter Soil Pollution Action Plan* issued by the State Council, the local *10-Chapter Soil Pollution Action Plan* has been published in 20 provinces, autonomous regions and municipalities such as Fujian, Anhui and so on, which highlighted the local characteristics of the soil environment in addition to decomposition and refining of the national standard. For example, Heilongjiang and Jilin list the black soil as one of the governance emphases. However, the current soil restoration capacity in China is still subject to technical and financial constraints. When it comes to technology, unlike governance of air pollution and water pollution, soil restoration is time-consuming and labor-intensive, and the process is more complicated. Plus the diverse soil types and numerous receptor (organism, water, air, human body, etc.) types, so the soil restoration technology is also complex. At present, many soil restoration technologies in China are still at the laboratory level and there is no effective technical system at present, as a result of which the key equipment still need to rely on imports. With regard to funds, the current soil restoration funds rely on government departments and land developers, but it is difficult for polluting enterprises and other subjects causing soil pollution to provide sufficient funds due to unclear property rights, limited scale and other reasons, and the principle of "whoever causes pollution is responsible for its treatment" is implemented with difficulty and is hard to be pushed forward, resulting in China's soil restoration capacity is difficult to gain rapid increase, still grim incremental pollution and common phenomenon of "toxic soil".

Shortcomings of the System Restrict Soil Restoration

"Pollution is visible while the system is invisible." At present, there are some gaps in the system in addition to technical difficulties for prevention and control of soil pollution, resulting in the bottleneck for enhancing prevention and control of soil pollution. Therefore, only by making up for the lack of system can we lead the way and escort for governance.

Firstly, pollution is incomputable and the existing data is not accurate. To have a clear understanding of soil pollution is the prerequisite and basis for all prevention and control. Since China has not yet established the farmland list and polluted land list to be classified by degree of pollution, supervision basis is insufficient. *Bulletin on National Soil Pollution Survey* released some test data of soil pollution, but the survey point is set to be more sparse, making the data difficult to accurately reflect the distribution and status of soil pollution, and even lack monitoring points and staffing in some areas. Therefore, we are in the urgent need to carry out a detailed survey on soil pollution with the focus on farmland and land for enterprises in major industries.

Secondly, the system has yet to be improved and the law guarantee needs to be strengthened. At present, China's soil pollution legal system has not yet been established, and relevant provisions are scattered in other environmental pollution laws and regulations. Legal system can prevent soil pollution violations, however, the current lack of legal absence makes it difficult for identifying legal responsibilities and subjects and so on, exacerbating the difficulty of dealing with cases in relation to soil pollution violations. At the same time, the law can also provide for the means of pollution control, and now less economic means is adopted in China, such as incentives to guide the market behavior. In the aspect of management of soil pollution, the United States has established a superfund site management system based on *Environmental Response, Compensation and Responsibility Law*, which includes accountability, restoration response and pollution compensation, that is, "whoever causes pollution is responsible for its treatment" and "whoever causes pollution shoulders the responsibility", serving as a strong deterrent to the polluters, which also becomes effective in pollution prevention and control in turn. In addition, due to legal basis is not fully realized, plus the seriously lagging soil environment monitoring, supervision and law enforcement and construction of risk early warning system in various places, it is difficult to exercise effective monitoring of soil environment within the area under administration.

Thirdly, the standard specification is not refined and the technical system needs to be improved. The existing soil environmental standard is not sound, making the current soil environmental pollution assessment and zoning lack appropriate basis. The existing *Soil Environment Quality Standard* (1995) is rough and less pertinent. Although the Ministry of Environmental Protection issued a series of standards for soil environment survey, monitoring, evaluation and restoration in construction land as a supplement, there are still many problems on the whole. First, too much emphasis is placed on establishment of soil standard and the reality of complex soil types in China is not fully taken into account, which fails to achieve adjusting measures to local conditions. Second, there are less pollutant projects, especially there are less types of organic pollutants. Third, heavy metal pollutant indicators focus on the total quantity control, but the soil pollution is mainly composed of the effective state, so the total quantity control cannot accurately reflect the actual impact of the soil. Fourth, establishment of the existing standard does not fully consider the toxic effect of soil on the biological receptors.

Soil Restoration Requires the Synergies of Policy and Market

Characteristics of the soil itself determine it easy to cause pollution and difficult to restore, which costs large input but is slow to take effect. It is foreseeable that the prevention and control of soil pollution will inevitably become an important campaign of the Chinese government to "declare war" against pollution. It will not only be a protracted war but also a tough fight. Therefore, we must pay attention to the establishment of a long-term mechanism, actively reduce the stock, tighten the control over the increment and strengthen seeking both temporary and permanent solutions to accelerate the continuous restoration and control of soil pollution.

Firstly, system improvement shall go ahead for prevention and control of soil pollution. First of all, we must combine with the implementation of the *10-Chapter Soil Pollution Action Plan* to speed up legislation for prevention and control of soil pollution. We shall adhere to the policy of "protection first, prevention orientation, prevention and treatment combination" and the principle of "classification, zoning, sub-target management" to establish the system for prevention and control of soil pollution that is in line with China's national conditions. In the prevention and control system, we shall set up the principle of fair and reasonable responsibility, and stipulate the corresponding legal responsibility according to the different legal liability subjects, and practice the principle of "whoever causes the pollution shoulders the responsibility". Second, we shall amend *Soil Environment Quality Standard* and other standards as soon as possible, such as polluted soil control and restoration, emission limit of heavy metal pollutants in major regional industries, analysis and testing methods of main pollutants, soil standard samples and other standards, as well as risk assessment, soil pollution control and restoration and other technical specifications, which can meet the needs of soil environment supervision. Third, we shall bring new ideas to supervision mode to enhance supervision capacity. For example, in the aspect of site supervision, we shall adopt the goal-oriented mode to promote the realization of the system project plan and site management of integration interaction between dynamic work strategy and real-time measurement technology dynamics according to the experience of New Jersey, thus to reduce the uncertainties in the process of site survey and restoration.

Secondly, we shall make market rules to stimulate the restoration vitality. Soil restoration also requires use to be good at utilizing economic means, focusing on the development of market rules to guide allocation of resources. When it comes to investment and financing, taxation and subsidies, we shall support the unified planning, construction and operation for environmental protection facilities in the region and achieve integrated management of environmental protection infrastructure. We shall also speed up the improvement of environmental investigation, analysis and testing, risk assessment, control and restoration engineering design and construction of the whole process of industrial chain development with city and county as the unit, get through the "blocked" links and give play to the role of "Internet+" in the entire industrial chain to promote soil, water, air pollution synergistic governance industry. We shall increase government's procurement of services through the government and social capital cooperation (PPP, Public-Private Partnership) mode to mobilize more social capital to participate in synergistic governance of the pollution. In addition, we shall play the guiding role of policy and development financial institutions to provide support for major pollution prevention and control projects for cross-regional and multi-media co-governance.

Thirdly, we shall call for the practical new technology that is in line with the current needs. China's research and development of polluted soil restoration technology shall provide technical support for solving problems such as soil pollution in farmland, soil pollution in industrial sites, soil pollution in mining areas and surroundings and soil pollution in ecologically sensitive wetland, which requires research and development to apply to different soil types and conditions, different soil utilization modes and the soil restoration technology with different pollution types and degrees. With regard to the agricultural soil or wetland soil which is polluted by heavy metals and organic matters, we need to focus on developing safer, low-cost and environmentally friendly biological restoration technology and material stabilization technology which can be put into large-scale application, and achieve production while restoration to safeguard safety of agricultural products and ecological safety. When it comes to soil in chemical, smelting and a variety of heavily-polluted sites which are abandoned by industrial and mining enterprises, we need to focus on developing site-pertinent restoration engineering technology which can meet the goal of safety and re-development and utilization, developing complete set of restoration technology and equipment with proprietary intellectual property rights and form systematic site soil restoration standard and technical specification to safeguard the safety and health of the living environment. With regard to the polluted soil in a variety of mining areas and tailings, we need to focus on researching the biostability and ecological engineering restoration technology to control water and soil loss and pollutant diffusion at this stage, thus to ensure the safety of the mining areas and the surrounding ecological environment and improve the ecological service value of the mining areas.

第二十五章　攻坚面源污染　共建美丽乡村

随着粮食产量的连增和棉油糖、果菜茶、肉蛋奶、水产品等农产品的充足供应，农业资源环境在支撑农业发展的同时，也长期透支了环境资源承载力，付出了巨大的生态代价，导致"红灯"频亮。由于化肥、农药、地膜等农业投入品过量使用以及秸秆、畜禽粪便和农田残膜等农业废弃物的不合理处置，造成农业面源污染加重。与此同时，垃圾围村、生活污水横流等农村环境污染问题日益凸显；加之工业污染由城市向农村转移，使得农村生态环境恶化，严重阻碍了美丽乡村的建设进程。因此，打好农业面源污染防治攻坚战、治理农村环境污染是推进现代农业建设、建设美丽乡村的重要任务，是"看得见山，望得见水，记得住乡愁"的基本要求。

面源污染形势严峻

一是农业面源污染来源多样化，资源综合利用程度低。我国农业面源污染来源主要可分为过量使用的农业投入品和缺乏高效综合利用的农业废弃物。我国化肥、农药生产和使用量均位居世界第一，但利用率却比世界发达国家低15%～20%。过量的化肥、农药、农膜在土壤和水体中大量残留，不仅影响了农业的可持续发展，而且还对食品安全构成了威胁。

畜禽粪便、秸秆、农膜等综合利用率低，成为重要的污染源之一。目前在全国，甚至一些规模化畜禽养殖场都没有配套污染治理设施，其产生的畜禽粪便、污水和恶臭气体对区域空气、地表水和地下水造成了严重污染，成为农村环境污染的主要因素之一。同时，我国是全球最大的水产养殖国，养殖产量约占全球的70%，面源污染使得水产养殖和环境保护之间的矛盾日益尖锐。

我国农用塑料薄膜覆盖面积大、消耗量多，每年使用量达260万吨，占世界消耗总量的60%以上。由于农膜较薄、难降解、回收率低，成为"白色污染"。我国每年产生秸秆9亿吨，秸秆就地焚烧不但浪费资源，而且污染空气。废弃的秸秆在田沟或河道中腐烂，进入水体造成严重的化学需氧量（COD）污染。虽然我国秸秆综合利用率达80%，但超过60%的秸秆被直接还田，或者制成肥料、饲料等，这些利用方法技术含量低，每吨秸秆的产值不足500元，这种低附加值的秸秆利用模式亟须转型升级。

二是农村生活垃圾、污水欠缺处理，环保基础设施薄弱、运行率低。我国约有4万个乡镇、60万个行政村和260多万个自然村，每年产生垃圾量大约为1.5亿吨。大部分乡镇由于资金匮乏，未配备专业的固定垃圾转运站和垃圾运输工具，生活垃圾乱扔、乱弃，"垃圾围村"现象非常普遍。据住建部统计数据显示，我国农村垃圾处理率只有50%左右。农村生活垃圾处理的普遍方式是简易填埋和露天焚烧，大量随意堆放的垃圾不仅破坏农村原本的自然生态面貌、浪费了资源，还经常引起害虫滋生，造成疾病传播和严重的环境污染。

我国乡镇污水处理设施建设同样滞后，河道、地下水和土壤污染问题突出，部分乡村"污水横流"。目前，我国有6.74亿农村人口，每天产生3000多万吨生活污水。随着生活用水、生产量的提高，农村生活污水、农业源废水污染物排放量还将持续处于高位。尽管污水处理技术日趋成熟，如厌氧—好氧—人工湿地或生态塘工艺等，但我国农村污水处理率很低，目前全国仅有10%左右的农村配套了污水处理设施，已建设施的运行率也不足10%；绝大多数农村污水未经处理而直接排入水体或渗入地下。

农村环境治理是场攻坚战

1. 农业面源污染的分散性、随机性和隐蔽性等特点加码治理难度。

面污染源十分分散，难以从源头加以杜绝。与集中排放的点污染源相比，面源污染"密度"远低于点源，且以扩散的方式发生，一般与气象变化相关，加之流域内土地利用状况、地形地貌、水文特征、气候、土壤类型等因素的差异,导致面源污染时空分布的异质性强。面源污染的分散性是面源污染治理的最大挑战。

面源污染的发生具有随机性和不确定性,污染控制措施却相对滞后。自然条件变化的随机性直接影响了面源污染发生的时间、区域及强度,致使面源污染的发生同样具有随机的特点。降雨量、温度、湿度等因素综合影响农作物的生产,进一步影响农药、化肥等化学制品的使用。面源污染的随机性使得其难以被及时监测,而相应的污染控制措施更难以被及时采用。

农业面源污染具有空间尺度和时间尺度的隐蔽性,为治理带来更大难度。农业面源污染由于排放的分散性导致其地理边界和空间位置不易识别,造成污染来源无法准确查实、及时追踪。同时,农业面源污染对生态环境产生影响一般是一个从量变到质变的过程,在量变的过程中具有隐蔽性和滞后性。例如,化肥中可能含有铬、镉、铅、汞等重金属,随着化肥的不断使用,这些成分会在土壤中不断积累,最终造成严重的环境污染。

2. 三大短板掣肘农村环境治理。

"观念旧"使得农村环境治理难以源头预防。受传统观念和习惯的长期影响,农村居民对面源污染的危害性认识不足,在农业生产中片面追求高产量,滥施滥用农药、化肥、农膜,随意丢弃、焚烧生产、生活废弃物,直排污水到河流水塘等。农村整体卫生意识欠缺,无论乡村干部还是普通群众,对于垃圾问题的重视程度远不及对农业增收的重视程度。

"钱不够"使农村环境治理力不从心。农业面源污染治理面广、量大、成本高,加上县乡基层财力有限,造成农村环保投入不足,垃圾处理的公共设施欠缺,填埋场的建设、垃圾工业化处理设备的使用在大部分农村都不普遍;很多农村没有配备专门的垃圾回收队伍和清扫队伍。滞后的基础设施建设与日益增长的污染负荷之间的矛盾日益突出,只停留在示范项目上的治理进程使得农村环境短期内难以得到根本治理。

"监管弱"使得农村环境治理未形成有效"组合拳"。我国针对农村环境治理的相应规定较为分散,缺乏统一性和协调性。基层政府的环保、住建、农业等职能部门在环境治理工作中职责交叉,阻碍了治理工作一体化的全域推进。另外，对于环保部门而言，尽管农村环境治理已提上议程，但不少地区乡镇一级的环保机构尚未建立，环保人才缺乏，而部分县级环保单位在应对工业、城市环境污染治理时已略显捉襟见肘，导致在农村环保治理上力不从心。

农村环境综合整治须久久为功

农业面源污染和农村环境的综合防治基础薄弱、工程复杂，与千家万户息息相关，需要不断提高认识、稳步推进、完善保障、久久为功。

一是明确落实农村环境各主体的责任。农业面源污染控制和农村环境治理的过程需要农民、企业、政府三方的共同参与，各方应当积极主动地承担责任，共同探索合理有效的农村环境治理机制。对于农民而言，应当不断提升"金山银山不如绿水青山"的保护意识，并积极运用到农业生产和农村生活中去，如使用优质易回收地膜、收集且不随意焚烧秸秆、做好垃圾分类及周围环境维护；同时应当积极配合有关单位缴纳相应的环保治污费用，如垃圾处理费等。对于企业而言，要充分承担起第三方治理的责任，提升资源化利用水平，提供农业面源污染控制、农村环境治理有关方案流程，保质保量地完成污染治理工作。对于政府而言，必须对农村环境问题肩负起总体把控、最终兜底的责任；针对不同地区制订不同的农业与农村污染治理办法，强化乡规民约保证；应当考虑到农村地区的资金困难，给予治污企业适当的财政补贴和税收优惠；积极推进乡镇一级环保机构的建立与环保人才的培养。

二是因地制宜、因势利导提升农村环境污染治理的技术水平。要确保农业面源污染、农村环境治理取得实效，必须将高效实用的治理技术作为改善农村生态环境的重要支撑。充分借鉴国内外先进技术，重点开发适合我国农村发展特点、居住特色以及高度集约化生产方式的技术。对于规模化养殖场粪便，应当在合理布局基础上建设配套处理设施，发展种养业废弃物资源化利用以及种养结合的循环农业，实现畜禽粪污资源化利用的目标。对于秸秆综合利用，应当将秸秆中的纤维素、半纤维素、木质素等成分加以分离，加工成具有较好市场前景的产品，提高秸秆综合利用水平。对于农村污水处理，应当注重少维护、抗冲击负荷、低能耗的农村污水处理需求，同时因地制宜，距离城市较近的农村应选择废水集中式处理，距离城市较远的农村应选择经济适用的生态处理工艺，充分考虑既有的水塘、滩地、沟渠，将设施建设、湿地保护、景观建设有机结合。

  三是发挥市场在农村治污中的资源配置作用。资金短缺是农村环境问题的症结所在，在我国农民收入普遍较低的背景下，解决资金问题的方法主要依靠政府的大力推动和市场的配合。应当持续推进生态循环农业试点、现代生态农业示范建设，不断培育农业面源污染治理市场主体试点，营造有利的市场环境，积极采用政府购买服务、政府和社会资本合作（Public-Private Partnership，PPP）、第三方治理等模式并加以创新，吸引更多的社会资金参与农业面源污染控制和农村环境治理。鼓励农村结合自身特色发展乡村旅游和绿色农业，让广大农民在脱贫致富的道路上守山护水，成为主人翁和践行者。

Chapter 25
Tackle Non-point Source Pollution to Build a Beautiful Rural China

With the continuous increase of grain output and the adequate supply of agricultural products such as cotton, edible oil, sugar, fruits, vegetables, tea, meat, eggs, milk and water products, etc., agricultural environment and resources have overdrawn the environmental resource carrying capacity while supporting the development of agriculture, resulting in an immense ecological cost and crisis. Overuse of fertilizers, pesticides, mulching films and other agricultural inputs and unreasonable disposal of straws, excrements of livestock and poultry, plastic film residues and other agricultural waste have led to serious agricultural non-point source pollution. Meanwhile, rural environmental pollution such as garbage and domestic wastewater has become prominent. In addition, industrial pollution is expanding from cities to the countryside, which has worsened rural ecological environment and hindered the process of building a beautiful rural China. Therefore, tackling agricultural non-point source pollution and improving rural environment is an important task in promoting agricultural modernization and a basic requirement for securing beautiful rural mountains and rivers which often makes people homesick.

Serious Non-point Pollution

Firstly, rural non-point source pollution has become diverse and comprehensive utilization of resources is insufficient. The sources of pollution include overuse of agricultural inputs and insufficient disposal of agricultural wastes. China ranks No. 1 in the world regarding the production and use of fertilizers and pesticides. However, the utilization rate is 15% ~ 20% lower than that in developed countries. Excessive residues of fertilizers, pesticides and plastic films in soil and water have not only hinders sustainable development of agriculture, but also posed great threats to food safety.

Because of low comprehensive utilization rate, excrements of livestock and poultry, straws and plastic films, etc. have become one of the most important sources of pollution. At present, even some big livestock and poultry farms lack pollution treatment facilities and the excrements, waste water and odorous gases generated have caused severe pollution to air, surface water and ground water in rural area. Meanwhile, China is the biggest aquaculture country and yields about 70% of the total aquaculture products in the word. Non-point pollution has resulted in increasing conflict between aquaculture and environmental protection.

Rural China consumes about 2.6 million tons of plastic films every year, accounting for over 60% of the total consumption in the world. Because these films are very thin and degradation-resistant, it is difficult to recycle them, resulting in white pollution. China produces 900 million tons of straws every year. Straw burning not only wastes resources but also pollutes air. Waste straw decay in field ditches or watercourses, causing severe COD pollution in the water. Although the comprehensive utilization rate of straws in China has reached 80%, over 60% of straws are returned to the field directly or made into fertilizers or fodders. These methods have low technical content and the economic value of one ton straws is less than 500 RMB. This model of utilizing straw with low added value should be upgraded urgently.

Secondly, environmental protection facilities in rural China are inadequate and inefficient regarding garbage and waste water treatment. Every year about 150 million tons of garbage is generated in the 600,000 administrative villages in over 40,000 towns. Most of these towns are short of funds and do not have special garbage collection centers or transportation vehicles, so domestic garbage is scattered everywhere. It is normal to see a village surrounded by garbage. According to statistics from Ministry of Housing and Urban-rural Development, only about 50% of rural garbage has been treated. The usual way of treating rural domestic garbage is simple landfilling and burning in open air. These randomly scattered garbage not only damages the natural landscape in rural area and waste resources, but also breeds a lot of pests, leading to the spread of diseases and severe environmental pollution.

Waste water treatment facilities in rural area also lag behind. Watercourse, ground water and soil pollution is severe. Waste water is flowing everywhere in some villages. At present, China has a rural population of 674 million and over 3,000 tons of domestic waste water is produced every day. With increasing consumption of water, the amount of domestic waste water and agricultural waste water will remain high. Although domestic waste water treatment technology is becoming increasingly mature, waste water treatment rate in rural China is very low. Currently only 10% of rural area is equipped with waste water treatment facilities. Only less than 10% of the completed facilities are in operation. Most of waste water in rural China is discharged directly to rivers or permeate into the ground without being treated.

Controlling Rural Pollution Is a Tough War

1. Rural non-point pollution exists in a scattering, random and hidden way, which makes it hard to treat.

It is difficult to eliminate non-point pollution from the source as it has very scattering sources. Compared with point source pollution with concentrated discharge, the density of non-point source pollution is much lower than that of point source pollution. It usually takes place in a diffusive way and often related to meteorological changes. In addition, differences in land use conditions, landforms, hydrological features, climate, soil types and other factors within the drainage basin lead to heterogeneous distribution of non-point source pollution in time and space. The dispersity poses the biggest challenges in non-point source pollution abatement.

Non-point source pollution takes place in a random and uncertain way while the treatment measures usually lag behind. The random changes of natural conditions directly affect the time, region and intensity of pollution. Rainfall, temperature, humidity and other factors affect the growth of crops, which will further impact the use of pesticides, fertilizers and other chemical products. The random nature makes it hard to monitor and even harder to adopt control measures in a timely manner.

Non-point source pollution in agriculture can be elusive in time and space, making it hard to control. It is difficult to identify the geographical boundary and spatial locations of the pollution because the pollution is very scattering, thus it is hard to confirm accurately the source of pollution. Meanwhile, the impact of pollution on environment is usually a process from quantitative to qualitative changes. At the beginning, the pollution can be hidden. For example, fertilizers may contain chromium, cadmium, lead, mercury and other heavy metals. With increasing use of fertilizers, these heavy metals will be accumulated in the soil, eventually causing severe environmental pollution.

2. Three Factors Hinder Rural Environmental Governance.

"Old Mindset" makes it hard to prevent rural pollution from the source. Influenced by traditional concepts and habits, rural residents are not aware of the danger of non-point source pollution. They use too many fertilizers, pesticides and plastic films in order to improve crop output. They throw away garbage at will, burn waste materials and discharge waste water directly into rivers and ponds, etc. Farmers as a whole lack the sense of hygiene and sanitation. Both village and township officials and ordinary farmers attach more importance to increasing agricultural yields than to controlling pollution.

"Lack of funds" makes it hard to improve rural environment. Controlling widespread non-point source pollution is costly and the local governments at county and township levels do not have enough financial strength, resulting in inadequate input into rural environmental protection. There are not enough public facilities for garbage treatment. It is not common to construct landfills and use industrial equipment to process garbage in rural area. There are no full time garbage collectors in the countryside. The backward facilities cannot meet the need of treating increasing pollution. The rural environment cannot be fundamentally improved within the short term with just some pilot projects on pollution control.

"Weak supervision" fails to facilitate synergy in rural environmental improvement. Regulations on rural environmental governance in China are fragmented. There is lack of coordination in grassroot-level government departments of environmental protection, housing and agriculture, etc. whose duties and functions are often overlapping, which has impeded pollution control in an integrated way. In addition, although the issue of rural environmental governance has been put into work agenda, environmental protection agencies at township levels have not been established and there is serious lack of environmental protection personnel. Environmental protection agencies at county level have already had too many problems to tackle regarding industrial and urban environmental pollution, so protecting rural environment far exceeds their capacity.

Long-term Efforts In Improving Rural Environment

The complicated project of tackling agricultural non-point source pollution and improving rural environment requires that long-term should be made to raise awareness and carry out work step by step because the project is vitally related to the interest of thousands of households and the infrastructure is weak.

Firstly, responsibilities of all the stakeholders in rural area should be clarified. The project requires the engagement of farmers, enterprises and government. All the stakeholders should actively undertake their responsibilities to find a reasonable and effective mechanism to improve rural environment. Farmers should continuously raise the awareness that clear water and green mountains are better than mountains of gold and silver. They should use high-quality and easily recycled plastic films, classify the garbage and protect the environment. They should collect straws rather than burning them at will. Meanwhile, they should pay relevant pollution control fees such as garbage disposal fees. The companies should shoulder the third-party responsibilities of improving resource utilization rate, recommending relevant plans on controlling agricultural non-point source pollution and improving rural environment and guarantee both the quality and quantity of pollution control work. The governments at all levels should shoulder the responsibilities of the final goalkeeper by designing different control methods in different regions, ensuring village regulations and agreements are obeyed, providing proper financial subsidies and tax incentives to pollution control enterprises in view of financial difficulties in rural area, promoting the establishment of environmental protection agencies at township level and training local talents in this field.

Secondly, improve technological level to control rural pollution in accordance with local conditions. Efficient and practical technologies should be used to improve rural environment in order to achieve tangible results. We should fully adopt advanced technologies at home and abroad and also develop new technologies which are in line with China's rural conditions. Supporting garbage treatment facilities should be built to treat excrements in some large livestock and poultry farms. We should develop circular agriculture that can utilize waste farming resources and integrate animal breeding with crop planting to achieve the goal of utilizing excrements and other waste resources in the farms. Regarding the comprehensive utilization of straws, we should separate cellulose, semi-cellulose, lignin and other elements which can be processed into products with good market prospects. Regarding rural waste water treatment, we should use the centralized method to treat rural waste water near the city. We should use affordable ecological treatment technologies to treat waste water in rural areas which are far from cities. We should integrate construction of waste treatment facilities, wetland protection and landscape design by making use of existing ponds, ditches and beaches.

Thirdly, fully display the role of market in allocating resources in rural pollution control. Shortage of funds is the root cause for rural environmental problems. In the context that the income of Chinese farmers is normally low, governments at all level should work with market forces to provide enough funding. We should promote pilot programs on ecological and circular agriculture and modern ecological agriculture, constantly nurture the development of market stakeholders in agricultural non-point pollution, build a favorable market environment, actively adopt and innovate the model of government purchasing services, PPP model and the third party governance model so as to attract more social capital to tackle rural pollution and improve rural environment. We should encourage the rural areas to develop tourism and green agriculture in line with their local features so that farmers can become practitioners in environmental protection when they shake off poverty and turn rich.

第二十六章　打造特色小镇要坚持生态优先

旅游小镇、科技小镇、金融小镇、文化小镇……近年来，随着中央一系列小（城）镇发展政策的出台，一大批功能各异的特色小镇纷纷涌现，呈现蓬勃发展态势。特色小镇依托当地自然禀赋、生态环境和历史文化积累等优势，聚焦特色产业，集聚发展要素，包含旅游休闲、健康养老、智能制造、商贸物流、美丽宜居等丰富主题。特色小镇不同于产业园区和风景区，也区别于行政建制镇，具有生态环境优美、产业定位独特、人文传统深厚、管理机制灵活创新的独特优势，成为推进新型城镇化建设、培育新动能、实现创新发展、绿色发展的新平台，充当着区域经济社会发展的新引擎。

特色小镇因何而兴

发展特色小镇是推进新型城镇化的新举措。城镇化的快速推进使得大大小小的城市迅速兴起，在集聚生产要素、带动经济发展、提升国际竞争力等方面发挥了重要作用，同时也带来了人口膨胀、交通拥堵、房价居高、重度雾霾等"城市病"。一边是不堪重负、亟待疏散的大城市，一边是序幕拉开、求贤若渴的特色小镇，供需反差之间隐含着巨大的人口流动势能，需要特色小镇主动出击、吸引人才。

正所谓小空间承载大战略。发展特色小镇是促进大中城市与小城镇协调发展的重要路径，凝聚了"创新、协调、绿色、开放、共享"的发展理念，是治理"城市病"、推进新型城镇化、实现城市可持续发展的新思路、新举措。特色小镇依托优美自然环境，挖掘地域传统文化，通过搭建创新创业平台，培育特色产业，促进要素集聚，发挥比较优势，实现差异化发展，为城镇建设和发展注入了创新动力。

发展特色小镇是统筹城乡发展、破解"三农"问题、建设美丽乡村的助推器。我国是一个经济社会发展不平衡的大国，存在着"城富乡贫，城新村旧"的现象。小镇是连接城市和农村的桥梁、发展现代农业的载体，也是要素流动的渠道。首先，小城镇大部分是由乡转变而来，普遍依托村庄建立，具有天然的沟通优势，是缩小城乡差距、实现城乡发展一体化的有力抓手；其次，特色小镇具有自身比较优势，在城镇建设和发展的某些领域能够体现人无我有、人有我强，其既能与都市经济融为一体，又能带动农业农村的发展；再次，小城镇的发展可以有效缓解农民就近就业问题，促进城乡公共服务均等化，使得村镇居民可以更快地分享城镇建设、产业成长、经济发展带来的好处。在贫困地区推进特色小镇建设，更有利于搭建脱贫平台，为精准扶贫提供载体。

发展特色小镇是推动生态文明建设和实现绿色发展的试验田。发展特色小镇是践行"绿水青山就是金山银山"的发展理念、走"生产、生活、生态"融合发展之路的有益探索。特色小镇，根在文化，既拥有现代化的生活，又保存乡土温情是特色小镇对时代问题的回答。特色小镇在发展自身特色的过程中，能够立足本地、放眼世界、扎根生活、连接社区，自觉承担起本地文化挖掘、保护、传承、创新的责任。

我国特色小镇所依托的乡镇地区大多尚未深度开发，自然环境优美，生态功能完善。特色小镇发展是以生态资源为本底、以绿色低碳循环技术为支撑、以绿色智慧产业为驱动、以创新体制机制为保障。特色小镇建设融入了生态文明、绿色发展理念，成为治疗"城市病"、改善农村环境质量、推进产业转型升级的有效尝试。

问题初现，特色小镇将何去何从

生态绿色制度供给不足。特色小镇建设是一项长期、复杂的系统工程，需要科学完善的顶层设计与制度保障。但我国目前仅有住建部、国家发改委、财政部"开展特色小镇培育"（2016），国家发改委"美丽特色小（城）镇建设"和"千企千镇工程"（2016），国家发改委、国家开发银行"开发性金融支持特色小（城）镇建设促进脱贫攻坚"（2017），体育总局"运动休闲特色小镇"（2017）等指导性文件。总体上看，我国特色小镇在生态文明建设、绿色发展等制度层面尚未系统设计规划；在国土、环保、文化、旅游、金融、卫生、水利、林业等领域政策支持力度不足；在规划建设标准、土地利用等方面缺乏法律制度保障和技术支撑。例如，特色小镇建设涉及农村土地使用权、农村宅基地、农村集体资产等使用、处分，而这些都没有充足的法律依据。

规划建设生态绿色理念不强。特色小镇的培育和发展关键在于发展定位与规划设计，应该坚持政府引导、市场主导，而不能一哄而上成为"政绩小镇"，必须抑制人为造"镇"的冲动；规划和建设应强化生态理念，防止成为大兴土木的"房地产小镇"；产业方向选择应坚持绿色标准，小镇不能变成一个筐，什么项目都往里装。地方政府在发展特色小镇过程中，应该与美丽乡村建设、田园综合体等其他政策实现协调发展、协同推进，避免大量"空心村""空心镇"的出现。

打造特色小镇要将生态绿色作为"必选项"

发展特色小镇要牢固树立生态优先绿色发展理念。特色小镇所要求的生态，是红线、底线，也是优势所在。要把生态文明理念和原则全面融入特色小镇建设的全过程和各领域，走出一条绿色、集约、智能、低碳的特色小镇建设之路。创新发展绿色经济，努力实现百姓富和生态美的有机统一，真正实现低碳生活、和谐生产、宜居生态。做大做强生态优势产业，有效增加生态产品服务供给，使生态优势和产业优势逐渐形成浑然一体、和谐统一的关系，为绿色发展奠定坚实的产业基础。

必须积极探索小城镇发展的路径，将生态绿色作为特色小镇的必选项。发展特色小镇要树立生态绿色理念，完善绿色政策顶层设计，建设运营全过程要强化生态绿色思维，产业选择须兼顾"特色"与"绿色"。

特色小镇是集产业发展、文化、健康养老、旅游休闲、体育运动等多种功能于一体的空间区域。小镇的开发建设要集约节约利用土地，空间布局与周边自然环境相协调，建设高度和密度相适宜，做到镇区环境优美、干净整洁，公园绿地贴近生活、贴近工作。

设计、建设、运营特色小镇要强化生态绿色思维。特色小镇强调的不是规模竞争，而是层次竞争，其目的是要形成在特定领域的持续影响力。要避免"立竿见影"的惯性思维，不要把小镇做成"摊大饼"，搞成新一轮的无序竞争。防止房地产开发和商业综合体在特色小镇总投资中占比过高、以特色小镇之名行房地产化之实，防止追求短期政绩和"面子工程"的现象。

特色小镇的发展，不在于磅礴浩大的体量建设，而在于小巧精致、舒适宜人的空间环境。在空间环境的塑造过程中要更加注重人的感受与体验。小镇的开发应当实施低冲击式的开发建设模式，在建设标准和功能定位上，着重突出生态、绿色、节能主题，用新材料、新技术打造低碳、舒适、宜居的环境；集约节约使用建设用地，推动规划区范围内的建设土地混合功能利用；倡导绿色交通模式，提高绿色出行比例；更加有效地节约使用水资源，倡导分散与集中相结合的水再生利用工程与水生态修复技术；以非工程措施应对雨洪威胁，保证小镇发展建设安全；大力推进生活垃圾分类、收集、资源化利用技术，减少噪音、污水、空气等污染。

特色小镇产业选择须兼顾"特色"与"绿色"。特色小镇须结合自身本底条件，审时度势，认真谋划产业选择。"特色"产业是特色小镇持续发展的"输血站"。其最关键的是产业业态的创新。要注重传统优势产业业态的提升和完善：制造业需要在"产品换代""机器换人""制造换法""商务换型""管理换脑"等"五换"业态上下功夫，积极探索业态创新的具体路径和方法。

要把绿色产业作为提升特色小镇竞争力的基本出发点。选择具有良好生态环境、经典文化传承和独特地形地貌的小镇，结合市场需求，发展文化旅游、健康养老、休闲度假、智能制造、"互联网+"、体育运动等业态。深度挖掘历史传统、民俗文化，彰显乡愁特色，结合现代生活方式，运用创意手段，融入生态绿色的元素，物质产品与精神产品并抓。发展生态循环农业，开发特色食品、健康食品，结合生态观光、农事体验、食品加工体验、餐饮制作体验等活动，推动体验式休闲度假。抓住承接产业梯次转移、疏解大城市功能的机遇，承接特定城市功能或特定产业。结合精准扶贫要求，统筹使用扶贫资金，实现贫困村产业创新与特色小镇建设的融合发展。

Chapter 26
Prioritize Ecology in Building Towns with Distinctive Features

A large number of towns with distinctive features and functions have sprung up and boomed such as tourism towns, technological towns, financial towns, cultural towns, etc. with the promulgation of policies on developing small towns by the central government. Relying on local natural endowments, ecological environment and cultural and historical inheritance and other advantages, these towns focus on special industries and concentrate development elements with rich themes including tourism and leisure, healthcare and elderly care, smart manufacturing, trade and logistics, livability, etc. Different from industrial parks, scenic spots and administrative towns, these towns boast unique advantages such as beautiful environment, unique industrial positioning, profound humanistic traditions and innovative and flexible management mechanisms and they have become new platforms in promoting new-type urbanization, fostering new momentum, achieving green and innovative development, acting as new engines for regional economic and social development.

How Special Towns Prosper?

Developing special towns is a new initiative to promote new type of urbanization. Rapid development of urbanization has led to fast rise of large and small cities. It has played an important role in concentrating productive factors, driving economic development and enhance international competitiveness. However, at the same time, it has resulted in many urban diseases such as population expansion, traffic congestion, high housing prices, severe smog, etc. On the one hand, some big cities are under too much pressure and the population has to shift to other areas. On the other hand, small cities and towns are eagerly looking for talents because they are at the initial stage of development. The imbalance between demand and supply means there is a great potential for population movement. These special towns or small cities need to take the initiatives to attract talents.

Small towns can uphold major strategies. Developing special towns is an important pathway to promoting the coordinated development between big and medium-sized cities and small towns, representing the concept of innovative, coordinated, green, open and shared development. It is a new approach and initiative to treating urban diseases, promoting new type of urbanization and achieving sustainable urban development. Taking advantage of beautiful natural environment and regional traditional culture, the special towns can build an innovative platform of entrepreneurship, foster industries with local color, promote the concentration of productive factors, display comparative advantages and achieve differentiated development, injecting innovative momentum into construction and development of small towns.

Developing special towns is the booster to balance urban and rural development, solving the issues of agriculture, farmers and rural areas and build a beautiful rural China. China is a big country with uneven social and economic development. The cities are rich and new while rural areas are poor and worn-out. Small towns are bridges connecting cities and the countryside, providing basis for modern agriculture and channels for the flow of productive factors. Firstly, most of the small towns are usually transformed from villages in the rural areas, so they are instrumental in bridging the gap between urban and rural areas and achieving integration of urban and rural development. Secondly, special towns have their own comparative advantages in many fields. They can both be integrated with urban economy and drive the development of agriculture in rural areas. Thirdly, small towns can create job opportunities for farmers and promote equalization of public services in urban and rural areas so that farmers in villages and small towns can share the benefits brought about by urbanization, industrial growth and economic development. Promoting the development of small towns in poor regions can build platforms to alleviate poverty to achieve targeted poverty reduction.

Developing special towns is the test field for promoting ecological civilization and achieving green development. Developing special towns is practicing the concept that blue mountains and clear water are mountains of gold and silver. It is a useful exploration of the pathway of integrating production, life and ecology. The root of these special towns is culture. Preserving rural sentiments while enjoying modern life is what special towns can offer in the contemporary era. In the course of development, special towns can integrate local life with the world, connect community members and automatically shoulder the responsibilities of exploring, preserving, inheriting and innovating local culture.

China's special towns are normally located in undeveloped places with beautiful natural environment and intact ecological functions. The development of special towns is based on ecological resources, supported by green and low carbon circular technologies, driven by green and smart industries and guaranteed by innovative mechanisms and institutions. Adopting the concepts of ecological civilization and green development, special towns become the effective initiative to treat urban diseases, improve rural environmental quality and promoting industrial transformation and upgrading.

In What Direction Will the Special Towns Take?

There is no adequate institutional arrangement for building ecological and green special towns. Construction of special towns is a long-term, complex and systematic project, which requires scientific and sophisticated top-level design and institutional guarantee. However, there are only several guiding documents in this area. For example, *Fostering Development of Specail Towns (2016)* issued by Ministry of Housing and Urban-Rural Development, National Development and Reform Commission (NDRC) and Ministry of Finance, *Construction of Beautiful Special Towns* and *Project of Building 1,000 Enterprises in 1,000 Towns (2016)* by NDRC, *Supporting the Construction of Special Towns to Alleviate Poverty with Financial Means* (2017) by NDRC and National Development Bank and *Building Special Towns for Sports and Leisure* (2017) by General Administration of Sports. In general, there is no systematic planning of special towns at the institutional level regarding ecological civilization construction and green development. Policy support is inadequate in the fields of environmental protection, culture, tourism, finance, health, water conservancy, forestry and land resources. There is lack of legal and institutional guarantee and technical support regarding planning and construction standards, land use, etc. For example, building special towns involves rural land use right and use and disposal of rural residence land and rural collective-owned assets, etc. However, there is no adequate legal basis at present.

There is a lack of green and ecological concepts in planning and construction of special towns. The key to the development of special towns lies in positioning, planning and design. We should stick to the principle of market-oriented development guided by the government and avoid the situation in which every region tries to develop special towns in large numbers just for political performance. We should suppress the impulse of creating special towns blindly. Ecological concepts should be reinforced in planning and construction to prevent building a large number of buildings in small towns just for the sake of real estate development. Green standards should be adopted in selecting industries. Development of small towns should not include unrelated projects. Local governments should integrate the development of small towns with construction of beautiful rural areas, rural complexes and other policies to avoid the emergence of many "hollow villages" and "hollow towns".

Ecological and Green Development Is a Must for Building Special Towns

We must actively explore the ways of developing small towns in line with the principle that ecological and green development is a must for building special towns. We should establish the concept of green and ecology, improve the top-level design on green policies. Green thinking should be emphasized in the whole process of construction and operation of special towns. While selecting industries for the special towns, we must give equal consideration to green and local features.

We must establish the concept of green development and ecology first when developing special towns. The ecological requirement of special towns is the red line, bottom line and the advantage. We must take the road of green, smart, low-carbon and intensive development and fully incorporate the concept and principle on ecological civilization into the whole process and every field of building special small towns. We must develop green economy in an innovative way and strive to preserve ecological beauty while enriching the farmers so as to achieve low carbon life, harmonious production and livable ecology. We must grow industries with ecological advantages bigger and stronger and effectively increase the supply of ecological products and services. Ecological and industrial advantages must be integrated harmoniously, laying a solid industrial foundation for green development.

Special towns can provide spaces integrating many different functions such as industrial development, culture, health, elderly care, tourism and leisure and sport activities. The development of small towns should fully utilizes the land by harmonizing spatial layout with surrounding natural environment with proper building height and intensity so that the towns are beautiful, clean and orderly and parks and green spaces are close to life and work.

Green and ecological thinking must be emphasized in designing, building and operating special small towns. The competitiveness of these towns lies in quality rather than size. The purpose is to build sustained impact in certain fields to avoid the inertial thinking of producing immediate effect. We must not develop the towns in a sprawling way and start a new round of disorderly competition. We must avoid the situation in which real estate development is conducted in the name of special towns and real estate development and commercial complex account for too high a proportion in the total investment of special towns. We must not seek short-term political achievements by building some shoddy projects.

The development of special towns should not focus on massive scale of construction, but on the creation of small, elegant, comfortable and livable spatial environment. Residents' feelings and experiences should be emphasized in the process of shaping the spatial environment. The development of small towns should adopt the construction mode of low impact. Regarding construction standards and functional positioning, themes like ecology, green and energy conservation should be highlighted and new materials and new technologies should be adopted to build a low-carbon, comfortable and livable environment. Construction land should be economized to promote mixed functional use of land in the planning zones. Green transportation mode should be advocated. Water resources should be used in a more efficient and economical way and water recycling technology and water ecological restoration technology should be advocated. Non-engineering methods should be used in response to flood threats to ensure the safety of small towns. Garbage classification, collection and recycling should be promoted to reduce noise, waste water and air pollution.

Green development and local color should be taken into account when industries are selected in special towns. Special towns should carefully plan what industries should be developed according to their local conditions. Industries with local characteristics should be the blood transfusion stations for the sustainable development of special towns. The key lies in innovation in industrial patterns. Traditional industries with competitiveness should be upgraded and improved. Manufacturing industries should focus on product upgrading, management improvement, technological innovation, business model change and staff training to actively explore innovative approaches and roadmaps.

Green industries should be the basic starting point to improve the competitiveness of these special towns. Small towns with excellent ecological environment, classic culture and unique landforms should be selected to develop industries such as cultural tourism, health and elderly care, leisure and vacation, smart manufacturing, internet plus and sports to meet market needs. We should further explore historical traditions and folk culture to highlight sentiments of nostalgia by integrating modern lifestyles, creative means and green and ecological elements and paying equal attention to both material and spiritual products. Circular and ecological agriculture should be focused to develop special and health food products and promote experience-based leisure and vocation activities by involving visitors in such activities as ecological sightseeing, farming experience, food processing and preparing local dishes. Small towns should seize the opportunities created by industrial transitioning and decentralizing functions of big cities to welcome some special industries. Poverty alleviation funds should be used in line with targeted poverty reduction so as to integrate industrial innovation in poor villages and the development of special towns.

第二十七章　呼唤动物福利时代

动物是生态环境的重要组成部分，动物福利水平是社会文明程度的重要标志。长期以来，我国对动物福利重视不够，特别是近年连续出现的"硫酸伤熊""高跟鞋虐猫""狗肉节"等虐待、残害动物事件，严重违背社会道德，影响了我国的国际形象。同时，动物福利的缺失也构成了国际贸易中一道新的贸易壁垒。其实，动物和人一样有感情、有情绪，只是它们无法用人类的语言表达见解。关注动物福利，就是关注人类自己。动物福利不仅可以使无数动物受益，也有助于提高人类自身的福祉。

动物保护运动莫衷一是的背后存在着不同利益之间的角逐与合作。产业界、动物保护界、科学界等之间及其各自内部都充斥着不同的诉求。如中医药界认为熊胆、犀牛角、虎骨等动物药疗效无法替代，主张在保护中华民族传统医学文化遗产的基础上进行野生动物利用；而动物保护界，即使是温和派，也希望通过一段时间的过渡而逐渐减少对野生动物的利用。即使在动物保护人士内部，理念也并不一致，人类中心主义者会以人类实用的观点来衡量动物保护的优先劣后和轻重缓急，而生态主义中心者则认为各种动物在生态系统中的作用难以估量，应当予以全面保护。

随着社会的发展，保护动物不断成为国际社会的基本共识。自20世纪70年代起，动物保护运动开始分为两派：动物权利者认为，动物与人一样，权利天赋，反对人类对动物的任何形式的占有和利用；动物福利者则主张以人为本人道地使用动物，同时给予动物应有的福利。作为世界动物卫生组织（World Organization for Animal Health，OIE）的一员，中国政府认可OIE规定的各项动物福利的原则和标准，包括动物享有不受饥渴的自由，生活环境舒适的自由，不受痛苦、伤害和疾病的自由，免受恐惧和表达天性的自由。

动物福利水平是社会文明的重要标志

1. 动物福利是现代文明的指数。从动物福利的定义看，提倡动物福利并非把动物置于人类之上，而是从生理、环境、卫生、心理和行为等五个方面保障动物的基本需要。善待动物其实是人类道德和责任向动物的延伸。人类对动物的轻贱残忍，往往会转化为对同类的邪恶；而善待动物、善待生命的人常常对整个社会心存善念。我们很难相信一个虐待动物的人会对社会抱有多大的善意，而关心动物的人往往对其他事物也充满慈悲。关注动物福利并非"多此一举"，也绝非伪善命题。促进动物福利，目的不仅是保护动物，更在于使人类远离暴力和伤害。当代中国不但不应该耻笑和排斥这种文明理念，反而应该正视和拥抱动物福利运动的发展趋势。

2. 动物福利立法是世界潮流。当前，全球已有100多个国家和地区建立了动物福利或反对虐待动物的法律。

许多发达国家都将动物福利与国际贸易紧密挂钩。2006年1月欧盟实施的《欧盟食品及饲料安全管理法规》特别加入了"善待动物"福利条款，向肉类企业提出了更高的要求。这意味着向欧盟出口的农产品，不但要符合欧盟食品安全的相关标准，还要延长食品安全管理链条。欧洲议会还通过了一项法令，要求在2009年之后欧盟各国的化妆品公司不得在动物身上进行化妆品的毒性或过敏性试验，欧盟也将禁止进口在动物身上进行过试验的化妆品。此外，欧盟委员会食品安全署还专门设有负责动物福利的部门。

3. 动物福利是中华文化的重要基因。动物福利的核心内涵是善待动物。虽然动物福利一词最早由西方提出，但其思想和理念并非"舶来品"。五千多年的中华文化传承着对生命尊重和道德伦理的推崇。儒家的"仁爱"思想，不只是对人，也包括对动物。孟子曰："君子之于禽兽也，见其生，不忍见其死。"古人很早就意识到，自然界生态平衡本就是人与动物和谐相处。虽然古人对动物的保护主要出于可持续利用动物资源、禁止奢侈挥霍风气等目的，与现代以动物权利或福利为出发点的保护理念存在区别，但不可否认的是，这些自古就有的动物保护思想和观念,对我们当今的动物保护理论与实践具有一定的启发和借鉴意义。

提高动物福利任重道远

我国自加入 OIE 以来,积极参与国际动物福利相关标准的起草和制定工作,在国内,先后修订和出台了《实验动物管理条例》等多项条例,并纳入了动物福利的内容。2016 年修订的《中华人民共和国野生动物保护法》(以下简称《野生动物保护法》)也明确规定"不得虐待野生动物"。我国还相继成立了一些旨在推广动物福利的行业协会,有效推动了动物福利理念的普及。但客观而言,我国的动物福利保护工作与国际社会相比仍存在较大差距。

1.公众对动物福利的认知存在误区和不足。首先,公众对"动物福利"一词陌生,感到荒谬可笑,甚至抵触。"动物福利"这一观念对国人来说确实"超前",因此常受到质疑非议。例如经常有人认为"人的福利还没有得到满足,何谈动物?"认为"农场动物终究要成腹中餐,赋予它们福利有何意义?"误认为动物福利与人类福利相冲突。如 2016 年某报《不要拥抱动物福利主义》一文认为:"动物福利主义满足了动物的福利,却降低了人类的福利,应小心对待。"事实上,"人的福利"主要指基本需求之外的额外的好处。而"动物福利"却是动物保障健康的基本需求。动物福利与人的福利是相辅相成的,提高动物福利,人类也是受益者。

其次,认为动物福利是一项需要高成本、高投入的工程,其实这是一种误解。实际上,农场动物的生存环境与人类食品安全息息相关。养殖户对畜禽进行福利养殖,主动改善饲养环境和落实无害化处理,能有效减少养殖污染;由于采取了更加尊重动物习性的喂养方法,会减少其生病率和死亡率;由于推行畜禽"人道屠宰"是基于动物的实际需要做出的选择,更有利于肉类产品质量的提高,从而保障人类食品安全,同时也是畜禽产品进入国际市场的必要措施。

2.动物福利立法长期滞后。目前,我国只有《野生动物保护法》和《中华人民共和国动物防疫法》(以下简称《动物防疫法》)、《中华人民共和国畜牧法》(以下简称《畜牧法》)、《生猪屠宰管理条例》《实验动物管理条例》等动物保护法律法规，缺少综合性的动物保护基本法，动物福利立法理念缺失，制约了动物保护整体水平的提升。野生动物保护法律体系相对完善，实验动物保护法律体系形式上初步具备，但实质上仍存在不系统、笼统概括、实效较差等问题。农场动物福利方面尚无国家法律及相应检测、评价与监管体系，但已有出台地方性的福利屠宰行业标准，如山东质监局发布的《肉鸡福利屠宰技术规范》。我国在伴侣动物保护方面，国家尚无专门立法，主管部门也有待明确，地方立法则主要涉及犬类动物。工作动物及娱乐动物的保护立法寥寥无几，基本集中于管理及防疫方面。

提高动物福利关键在于理念与法制

动物福利不是"假慈悲"，关注动物福利就是关注人类自己。逐步走向国际舞台聚光灯下的中国亟须将"动物福利"纳入到生态文明的体系中来。

1. 加强公众参与，提高公众对动物福利的认知水平。善待动物是中华文明的优秀传统，是社会主义核心价值观的内在要求。要提升国民普遍的动物保护意识，普及"动物福利"理念。要让农场主认识到良好的动物福利可以提高动物的免疫力，实施更加人性化的福利养殖技术才能让动物更健康，让食品更安全，从而让养殖场持续盈利。积极引导培育一批民间动物保护公益组织，加强公益组织的专业化水平和自身能力建设，正确引导公益组织依法、有序、高效地为我国动物保护事业发挥积极作用。

2. 完善动物保护法律体系、增强动物福利理念。目前，世界各国的动物保护立法大都基于动物福利保护和禁止虐待动物。动物福利立法是在保证并推进人类可持续发展的目标下去关怀、爱护动物，对动物的生存环境等进行优化，以达到对动物安全、人道的利用。

尽快制定适合我国国情的《动物福利法》，提高动物福利水平。一要在《野生动物保护法》的基础上，严格落实"不得虐待野生动物"等要求，加强对于圈养野生动物的监管和福利保护。二要针对农场动物福利的适用范围、产业链环节控制、监管机制、进出口福利保护、法律责任等做出明确规定。三要科学、合理、人道地使用实验动物。修订《实验动物管理条例》等相关法律法规、技术规范，落实"减少、替代和优化"的3R原则，提高实验动物福利水平，解决医药品、化妆品等因实验动物测试而造成的贸易壁垒问题。四要充分考虑国情，在尊重文化差异的前提下，确立伴侣动物登记许可、饮食居所、医疗免疫、流浪收容、陪伴、繁育等针对不同种类伴侣动物的详细保护要求。五要明确工作动物的工作强度、时间、年限、环境等福利规定。六要严格保证娱乐动物在被饲养、训练表演等过程中的身心健康，同时须逐步减少娱乐动物的数量。

Chapter 27
The Era of Animal Welfare

Animals are an important part of the ecological environment and level of animal welfare is an important indicator of the degree of social civilization. For a long time, China did not attach enough attention to animal welfare. In particular, some incidents related to animal abuse and mutilation, such as injuring bear with sulfuric acid, abusing a cat with high heels, dog meat festival, etc., which took place in the last few years, have seriously violated social morality and damaged China's international image. Meanwhile, absence of animal welfare also constitutes a new trade barrier in international trade. In fact, animals, like human beings, have feelings and emotions, but they cannot express their views with human languages. Paying attention to animal welfare is showing care for the mankind. Animal welfare not only benefits countless animals, but also helps to improve human wellbeing.

Behind different views on animal protection movements there exist conflicts of interests. Industries, animal protection circle and science community all have different interests. For example, traditional Chinese medicine industry believes that the efficacy of medicine made from bear bile, rhinoceros horns, tiger bones and other animal parts are irreplaceable, so they maintain wild animals could be used on the basis of protecting traditional Chinese medicinal culture and heritage. However, animal protectionists, even the moderate ones, hope to gradually reduce the use of wild animals within a period of time. Animal protectionists also have different ideas among themselves. People who believe in anthropocentrism would adopt a pragmatic view to measure the priority and urgency of animal protection from the perspective of mankind while people who believe in eco-centrism hold that animals should be protected because they play an immeasurable role in the ecosystem.

With the development of society, protecting animals has gradually become the basic consensus of the international community. Since 1970s, animal protection movement began to be divided into two factions: animal right advocates hold that animals have innate rights like human beings, so they oppose ownership or use of animals in any form. Animal welfare proponents maintain that animals can be used in a humane way and that at the same time, animals should enjoy animal welfare. As a member of the World Organization for Animal Health (OIE), the Chinese government recognizes the principles and standards on animal welfare set by OIE including animals enjoying freedom from thirst and hunger, freedom from discomfort, freedom from pain, injury, and disease, freedom to express most normal behavior and freedom from fear and distress.

Animal Welfare Level Is an Important Symbol of Social Civilization

1. Animal welfare is an index of modern civilization. From the definition of animal welfare, promotion of animal welfare is not to put animals above human beings, but rather to ensure the basic needs of animals regarding physiology, environment, health, psychology and behavior. Treating animals in a humane way is an extension of human morality and responsibilities to animals. Mankind's meanness and cruelty to animals are often transformed to wickedness towards human beings. Those people who treat animals well are usually compassionate to the whole society. We do not believe a person who abuses animals is benevolent while a person who care about animals is always filled with mercy. Concern about animal welfare is not superfluous or hypocritical. The purpose of promoting animal welfare is not only protecting animals, but also protecting human beings from violence and injury. Modern China should not sneer at or reject this civilized concept. On the contrary, we should face squarely and embrace the trend of animal welfare movement.

2. Animal welfare legislation is the world trend. At present, more than 100 countries and regions have enacted laws on animal welfare or opposing animal abuses.

Many developed countries have closely linked animal welfare with international trade. *EU Regulation on Food and Feed Safety Management* implemented in January 2006 specifically added one provision on animal welfare, which put higher requirement on meat companies. This means that agricultural products exported to the EU should meet the EU food safety standards. Furthermore, animal welfare should be extended to food safety management chain. European Parliament also passed a decree calling for cosmetics companies in EU not to carry out cosmetic or allergic tests on animals after 2009. EU would also ban the import of cosmetics that had been tested on animals. In addition, European Food Safety Authority has also set up a dedicated department responsible for animal welfare.

3. Animal welfare is an important gene of Chinese culture. The core connotation of animal welfare is to treat animals humanely. Although the concept of animal welfare was first proposed by the West, this idea is not exotic. Chinese culture with a history of 5,000 years has inherited respect for life and moral ethics. The concept of benevolence advocated by Confucianism covers both mankind and animals. Mencius said, "A gentleman is happy to see animals live and he cannot bear to see them die." Ancient Chinese realized that ecological balance in nature means harmonious co-existence between man and nature. Although ancient people protected animals for the purposes of utilizing animal resources in a sustainable way and banning extravagant lifestyles, which is different from the modern protection concept whose starting point is on animal rights or welfare, there is no denying that these ideas and concepts on animal protection in ancient times can be instructive to today's theory and practice on animal protection.

There Is Still a Long Way to Go Regarding Improving Animal Welfare

Since joining the OIE, China has been actively involved in the drafting and formulation of standards related to international animal welfare. At the national level, China has revised and promulgated *Regulations on the Administration of Experimental Animals of the People's Republic of China* and many others, which include the contents of animal welfare. *Law on Wildlife Protection* amended in 2016 expressly stipulates that ill-treatment of wild animals is prohibited. China has also set up a number of industry associations to promote animal welfare, effectively promoting the popularity of animal welfare concept. But objectively speaking, there is still a big gap between China and international community regarding animal protection.

1. There is a misunderstanding and lack of public awareness of animal welfare. First of all, being unfamiliar with the concept of animal welfare, the public feel ridiculous and even challenge or oppose this concept. They think the concept is far too advanced for the Chinese people. For example, some people think mankind's welfare has not been ensured, let alone animal welfare. They also believe it is meaningless to ensure animal welfare because eventually farm animals will become delicious food in people's stomachs. They mistakenly believe that animal welfare will conflict with human welfare. For instance, an article entitled *Never Embrace Animal Welfarism* holds that "animal welfarism improves animals' welfare, but decreases human welfare, so we should take the concept cautiously". In fact, human welfare mainly refers to extra benefits beyond basic needs while animal welfare means ensuring the basic needs of health of animals. Animal welfare and human welfare are complementary and mankind can benefit from improving animal welfare.

Secondly, there is a misunderstanding that improving animal welfare is a high-cost and a high-input project. In fact, the living environment of farm animals is closely related to human food safety. Farmers paying attention to animal welfare will actively improve breeding environment and the implement harmless treatment so as to effectively reduce pollution. Because they adopt feeding methods that respect animals' habits, morbidity and mortality of animals will be reduced. Because they implement measures of slaughtering animals in a humanely way, which is based on the practical needs of animals, the meat quality can be improved to ensure human food safety. Meanwhile, it is also an necessary measure for the animal products to enter international market.

2. Animal welfare legislation has long lagged behind. At present, China has only made the following laws and regulations on animal protection: *Law on Wildlife Protection, Animal Epidemic Prevention Law, Animal Husbandry Law, Pig Slaughtering Act, Regulation on Experimented Animals*, etc. and there is no

comprehensive animal protection law. The lack of legislative concept on animal welfare has restricted animal protection. The legal system on protecting wild animals is comparatively sound. Legal framework on protecting experimented animals has been preliminarily established, yet in essence, it is too general, unsystematic and ineffective. There is no national law or corresponding testing, evaluation and supervision system on farm animal welfare. However, local industrial standards on slaughtering have been issued in some regions, e.g., Technical Standard on Slaughtering Chicken issued by Shandong Provincial Quality Supervision Bureau. China has not enacted any laws on companion animal protection. Competent government departments need to be clarified. Local legislation mainly covers dogs regarding management and epidemic prevention. There is basically no law covering work animals or recreational animals.

The Key to Improving Animal Welfare Lies in Legal Protection and Change of Mindset

Animal welfare is not false compassion and caring about animal welfare is caring about human beings themselves. China needs to include animal welfare into the system of ecological civilization since China has gradually stepped into international limelight.

1. We should emphasize public participation and raise public awareness of animal welfare. Treating animals humanely is the fine tradition of the Chinese civilization. It is the inherent requirement of the core socialist values. We should raise public awareness of animal protection and popularize the concept of animal welfare. Farmers should realize that good animal welfare can help to improve animal's immunity and that adopting humane breeding technologies can make animals healthier and food safer, thus making more profit. We should foster a large number of non-governmental animal protection organizations, enhance their professionalism and capacity building and guide these NGOs to develop in a legal, orderly and efficient manner so as to play a positive role in China's cause on animal protection.

2. We should improve the legal system on animal protection and enhance the concept of animal welfare. At present, animal protection legislation in the world is largely based on animal welfare protection and prohibition of cruelty against animals. The purpose of enacting laws on animal welfare is to care about and love animals, optimize the living environment of animals so as to ensure safe and humane use of animal resources in line with the goal of promoting human sustainable development.

We should enact *Animal Welfare Law* in line with China's condition to improve animal welfare. Firstly, on the basis of *Law on Wildlife Protection*, we must strictly implement the requirement of no abuse of wild animals and strengthen the supervision and welfare protection of captive wild animals. Secondly, the scope of the application of farm animal welfare, industrial chain link control, regulatory mechanisms, import and export welfare protection and legal responsibilities should be defined. Thirdly we should experiment with animals in a scientific, reasonable and humane way. Relevant laws, regulations and technical specifications should be revised to implement the 3R principle of "reduction, replacement and refinement", improve the level of welfare of experimental animals, solve the problems of trade barriers caused by laboratory tests related to pharmaceuticals and cosmetics. Fourthly, we must fully take national conditions into consideration and put forward detailed requirements on the protection of companion animals regarding registration permits, food and accommodation, medical care and immunity, vagrant animals sheltering, accompaniment, breeding and other aspects on the precondition of respecting cultural differences. Fifthly, we should clarify the stipulations on welfare of working animals, such as work intensity, work time, years and environment, etc. Sixthly, we must ensure physical and mental health of recreational animals in the course of breeding, training and performance. Meanwhile, we should gradually reduce the number of recreational animals.

第二十八章　维护生态安全　推进永续发展

自国家出现起，生态安全就与国家的生存发展紧密相关。夏禹以治水而平定九州，楼兰因缺水而亡于瀚海，兴亡之间说明只有维护生态安全，才能使国家存续进而繁荣昌盛。生态安全作为国家安全的重要组成部分，已经成为关系人民福祉和民族未来的大事；将生态安全纳入国家安全体系，是推进国家治理体系和治理能力现代化、实现国家长治久安的现实要求；对于促进经济社会可持续发展，加快生态文明建设具有重要意义和深远影响。必须以全球视野、长远眼光、系统思维看待生态安全问题，厘清当前生态安全领域的问题与挑战。

生态安全是国家安全的根底沃土

生态安全，是指一个国家所具有的自然环境条件，不会危害或威胁该国家生态—经济—社会系统运行和发展的状态。生态安全既是经济、社会、政治等其他传统安全的基石，又受到传统安全的影响和制约。生态一旦遭到破坏，将使人们丧失大量适宜生存的空间和资源，甚至由此产生"多米诺骨牌效应"，随之击溃经济、摧垮政权、荼毒生灵、毁灭文明。随着资源短缺、气候变化及各国角力日益严峻，生态安全问题早已超越自然科学领域，逐渐演变为关系国家生存发展的政治议题，各国纷纷借保护生态为本国权益寻求有利的证据，寻求世界生态舞台上的道义制高点，因此生态安全始终要以国家作为出发点和立足点。

生态安全是事关国家主权与生存发展的国家安全重要组成部分。现代国家安全的概念不再局限于国防安全、军事安全等传统因素，而愈来愈多地包括土壤、水、森林、气候等作为物质生活基础的生态环境。生态安全的危险后果具有一定的潜伏期，渐进累积达到临界阈值时就会突然爆发。目前我国的生态环境形势不甚乐观,生态恶化的趋势尚未得到根本扭转,国家生态安全面临前所未有的挑战:环境污染严重,危害人民群众的身体健康;生态赤字扩大,削弱经济社会的发展基础;物种资源破坏,制约人类社会的长远发展;自然灾害频发,造成生命财产的重大损失。

生态安全是经济安全、社会安全、文化安全、政治安全、军事安全的载体和基础。从生态安全与经济安全的关系来看,自然资源是经济系统的投入要素,环境是经济系统得以运行的必要条件,只有在经济安全得以保障的前提下,国家才有足够的资金投入生态治理进而保障生态安全。从生态安全与社会安全的关系来看,近年来生态环境问题社会化形势愈发严峻,环境群体性事件高发,生态安全的破坏,必然影响社会稳定,危及国家安全。从生态安全与文化安全的关系来看,"竭泽而渔"既是生态危机,更是文化危机,个人主义价值观、享乐主义发展观、消费主义生活观是可持续发展观的宿敌,维护生态安全必须纠正长期以来"重经济轻生态"的观念。生态安全与政治安全的关系往往会上升到国际层面,生态安全可能引发外交摩擦与国际争端,而生态—政治安全矛盾的激化,甚至会爆发军事冲突。由此可见,生态安全与其他安全不仅相互依存、相互交织、相互渗透、相互牵制,甚至在一定条件下会发生相互转化。

生态安全问题错综复杂尚待条分缕析

维护生态安全是一个复杂的系统工程,需要对当前我国面临的生态问题进行全方位的把握。总的来看,当前生态安全面临着资源供应趋紧、环境污染严峻、生态安全管理部门职能分散等问题,而主权问题又增加了生态安全的复杂性。

第一,资源供应趋紧,环境问题告急。从总体上看,我国资源人均占有量低。水资源、耕地资源、森林资源人均占有量仅为世界人均占有量的28%、33%、20%;能源、矿产资源对外依存度高,如石油对外依存度达59%,铁、铜对外依存度达60%以上。而且,资源的开发利用方式粗放,浪费严重。目前我国水能的利用率只有6%左右，煤矿采选回收率仅有40%左右。环境污染涉及多个领域，污染程度不容乐观。生物多样性面临挑战，当前34450种高等植物中受威胁物种占10.5%，4357种脊椎动物（不含海洋鱼类）中，受威胁物种占21.4%。大气、水、土壤等多个圈层环境质量形势严峻。大气方面，2016年338个地级及以上城市平均超标天数比例21.2%，广大中东部地区雾霾天气频繁，南方地区酸雨多见。水体方面，2016年6124个地下水监测点中较差及以下的监测点比例高达60.1%。土壤方面，全国土壤总超标率达16.1%，耕地土壤环境质量堪忧，工矿业废弃地土壤环境问题突出。自然生态系统逐渐呈现出由结构性破坏向功能性紊乱的方向发展，因此保障生态安全压力颇高。

第二，管理职能分散，缺乏国家统一决策。目前，我国政府采取按生态和资源要素分工的部门管理模式，生态安全管理职能分散在各个部门，在国家层面缺乏统一决策、统一监管的体制机制，造成公共利益和部门行业利益的冲突。自然资源、环境生态保护职能分散在国土、水利、农业、林业等部门，管理职能的分散性与生态系统的完整性易产生冲突，同时生态环境法律政策仍存在不协调的现象。

第三，主权问题使得生态安全盘根错节更为复杂。我国有不少一衣带水的邻邦，如果与周边国家关于国土、水资源等相关争议处理不当，必将会对政治、外交造成不利影响。环境问题引发主权国家之间的争议较为间接，却不可忽视。沙尘暴、雾霾等跨国污染问题正逐渐受到各国的高度重视。因此，生态安全问题难免同主权国家的利益交织在一起，增添生态安全问题的复杂性。

生态安全靠制度、需技术、协全球

为了维护生态安全，保障国家的长治久安和民族的永续发展，需要构建生态安全体系，排除威胁生态安全的一切因素。

第一，推进生态安全治理体系和治理能力现代化。生态安全是一项重大的系统性工程,必须在国家层面注重顶层设计。推进生态安全法治建设。在现有各类法律法规基础上,立足国家生态安全需求,厘清生态安全与资源、环境保护的关系,健全具有中国特色的生态安全法律体系。加快生态安全体制机制建设,整合相关的组织机构,明确各部门职责。国家层面要建立有效的监督考核与问责机制,确保国家生态安全战略实施的效果。各级党委和政府应对本辖区的生态安全状况负责,将国家生态安全工作纳入国民经济和社会发展规划。完善民主监督制度。开展生态安全教育,提高广大干部群众的生态安全意识,主动地监督危害国家生态安全的行为,形成良好的社会法治环境。

第二,整合重大工程,构建安全屏障。整合现有各类重大工程,构建生态保护、经济发展和民生改善的协调联动机制,用制度保护生态环境。划定生态保护红线,按照生态系统完整性原则和主体功能区定位,优化国土空间开发格局,理顺保护与发展的关系,改善和提高生态系统服务功能,构建结构完整、功能稳定的生态安全格局,从而维护国家生态安全。

第三,完善技术支撑,观测、监管、预警系统一体化。利用卫星遥感技术等手段,建设生态遥感观测体系,定期对指定区域进行遥感监测和实地核查。充分挖掘和运用大数据,综合采用空间分析、信息集成、"互联网+"等技术,构建国家生态安全综合数据库,通过对生态安全现状及动态的分析评估,完成监测工作,并预测未来生态安全的情势及时空分布。建立国家生态安全的监测、预警系统,及时掌握国家生态安全的现状和变化趋势,建立警情评估、发布与应对平台,充分保障我国生态安全,为国家提供相关的决策依据。制定国家生态安全的评价标准,对国家生态安全状况进行总体评价,并定期发布国家生态安全状况,让全社会直观、形象地了解我国的生态环境状况,提高人民群众对生态环境的关注度。根据当地生态环境状况,建立和完善适应当地的生态安全预警和防护体系。

第四,推进国际协调机制建设,于全球生态治理中担当大任。从全球层面来看,对生态安全的关切点集中于气候变化和跨国大气污染等方面,可以在联合国等机构内设置专门机构,在国际法的框架里负责全球生态安全法规的制定和监督实施；可以借助联合国环境署、世界自然基金会等国际组织协调各国的生态安全争议问题。从地区层面来看，对生态安全的关切点更为广泛，包括水资源、矿产资源的争夺及水污染和大气污染的争议等，可以借助地区组织及国家间的对话磋商机制来协调生态安全问题。从我国实际情况出发，应引领"一带一路"沿线各国共建绿色丝绸之路。以生态农业、生态工业、生态旅游等绿色经济为核心，以"一带一路"为契机，深化中国环保国际化。借此促进国际交流，提升环保实力，维护生态安全，彰显大国风采。

Chapter 28
Maintain Ecological Security and Promote Sustainable Development

Since the emergence of countries, ecological security has been closely related to the survival and development of these countries. Yu, an official in Xia Kingdom of ancient China, maintained social stability by controlling floods. Loulan, an ancient kingdom of China, disappeared in history for lack of water. The rise and fall of these countries demonstrate that a country can survival and prosper only when she maintains ecological security. As an important part of national security, ecological security has become a major factor in relation to the well-being of people and the future of a nation. Incorporating ecological security into the system of national security is a realistic requirement for promoting the modernization of national governance systems and governance capabilities and achieving long-term national stability. The concept has great significance and profound impact on promoting economic and social sustainable development and accelerating ecological progress. We should look at ecological security systematically with a global and long-term vision and understand the issues and challenges in the field of ecological security.

Ecological Security Is the Foundation of National Security

Ecological security means that a country's natural environmental conditions do not jeopardize or threaten the operation and development of its ecological, economic and social systems. Ecological security is the cornerstone of traditional security like economic, social and political security and it is also influenced and restrained by traditional security. Once ecology gets damaged, people will be deprived of considerable space and resources suitable for survival and even a domino effect will follow, resulting in economic meltdown, regime collapse, loss of lives and destruction of civilization. With the shortage of resources, climate change and increasing national competitions, ecological security have long been beyond the field of natural sciences and gradually evolved into a political issue related to the survival and development of a country. Countries in the world have tried to seek favorable evidences for their national interests on the pretext of protecting ecology and occupy the moral high ground in the global stage of ecology. Therefore, the starting point and foothold for ecological security should be the national interests.

Ecological security is an important part of national security related to national sovereignty, survival and development. The concept of national security in the modern sense is not just limited to military security and national defense security and other traditional factors, but rather it increasingly includes ecological environment consisting of soil, water, forest and climate, etc. as the foundation for material life. There is a certain incubation period before dangerous consequences of ecological security gradually reach a critical threshold and suddenly break out. At present, China's ecological environment is not optimistic, the trend of ecological deterioration has not yet been fundamentally reversed and national ecological security is facing unprecedented challenges: serious environmental pollution endangering people's health; widening ecological deficit weakening economic and social development; damaged species resources restricting the long-term development of human society and frequent natural disasters resulting in significant loss of life and property.

Ecological security is the basis and carrier of economic, social, cultural, political and military security. Regarding the relationship between ecological security and economic security, natural resources are the input factors of economic systems. A sound environment is a necessary condition for the operation of economic systems. Only when economic security is guaranteed, can a country have sufficient money to be invested into ecological governance to ensure ecological security. Regarding the relationship between ecological security and social security, ecological and environmental issues have become increasingly serious, causing social unrest in the last few years. The damage to ecological security will inevitably affect social stability and endanger national security. Regarding the relationship between ecological security and cultural security, killing the goose that lays golden eggs is both an ecological crisis and a cultural crisis. Such values as individualism, hedonism and consumerism are the old enemies of sustainable development. The concept of prioritizing economic development over ecology practiced for a long time should be corrected in order to maintain ecological security. The relationship between ecological security and political security has always been internationalized. Ecological security can trigger diplomatic frictions and international disputes while escalating contradictions between ecological security and political security can even lead to military conflicts. Therefore, ecological security and other types of traditional securities are interdependent, intertwined, mutually penetrating and mutually restricting. Even they can be mutually changeable under certain conditions.

Complex Issues on Ecological Security Remain to Be Further Analyzed

Maintaining ecological security is a complex and systematic project, requiring us to have a comprehensive understanding of the current ecological problems facing China. Generally speaking, these problems include increasing tight supply of resources, severe environmental pollution and fragmented responsibilities and duties in different government departments in charge. Furthermore, the issue of sovereignty increases the complexity of economic security.

Firstly, supply of resources has become increasingly tight and environmental problems have become a growing concern. On the whole, China's per capital resources are very limited. Her resources of water, arable land and forestry per capita account for respectively 28%, 33% and 20% of the world averages. China depends a lot on importation regarding mineral and energy resources. For example, China's foreign dependency ratio for oil reaches 59% and over 60% for iron and copper. Furthermore, China's development and utilization of resources is extensive and wasteful. At present, the water utilization rate is only about 6% and coal mining recovery rate is only 40%. Environmental pollution in many fields is very severe.

Biodiversity faces many challenges. 10.5% of the 34,450 higher plants are under threat and 21.4% of the 4,357 vertebrates (excluding marine fishes) are under threat. Environmental quality of air, water, soil and many other spheres is grim. Air quality in 338 cities is sub-standard for 21.2% of the days in 2016. Smoggy days are frequent in central and eastern regions. Acid rain is common in South China. Groundwater quality in 60.1% of the 6,124 water monitoring sites is poor in 2016. 16.1% of national soil is substandard. The soil quality of arable land is low. Environmental issues related to soil quality in industrial and mining wasteland are very serious. The ecosystem is developing from structural damages to functional disorders. Therefore, ensuring ecological security is under great pressure.

Secondly, the management functions are fragmented, resulting in lack of unified national decision-making. At present, China adopts the ministry-based management mode according to the division of ecological factor and resource factors. The management functions for ecological security are scattered in different government departments, resulting in an institutional management mechanism that lacks unified decision-making and regulation at the national level and causing conflicts between public interests and departmental and industries interests. The functions protecting natural resources and ecology are scattered in different departments like land and resources, water conservancy, agriculture, forestry, etc. Conflicts will easily arise between the fragmentation of management functions and the integrity of ecological system and there is a lack of coherence in laws and regulations on ecology and environment.

Thirdly, the issue of sovereignty makes ecological security more complex. China has many neighboring countries. Improper handling of disputes related to borders and water resources, etc. can have an adverse impact on political and diplomatic relations. The disputes over environmental issues between sovereignty countries are indirect, yet they cannot be ignored. All countries in the world attach great importance on cross-border pollutions such as sandstorms and smog. Therefore, the issue of ecological security may inevitably be intertwined with interests of sovereignty states, increasing the complexity of ecological security.

Ecological Security Requires Institutional Arrangement, Technology and Global Coordination

An ecological security system should be established to eliminate all factors threatening ecological security in order to maintain ecological security and protect long-term stability and sustainable development of a nation.

Firstly, we should promote ecological security management system and modernization of governance capabilities. Ecological security is a major systematic project that must focus on top-level design at the national level. We should promote the rule of law. Based on the existing laws and regulations, we should focus on national needs for ecological security, clarify the relationship between ecological security and environmental protection and improve ecological security-related legal system with Chinese characteristics. We should enhance institutional arrangement related to ecological security, consolidate relevant organizations and identify duties and responsibilities of different departments. An effective supervision, assessment and accountability mechanism should be established at the national level to ensure the effect of implementing a national strategy on ecological security. Party committees and governments at all levels should be responsible for ecological security in their respective regions and incorporate relevant work into plans on economic and social development. Democratic supervision system should be improved and education on ecological education should be conducted to raise public awareness of ecological security, actively monitor behaviors of endangering national ecological security and build a favorable social and legal environment.

Secondly, we should consolidate major projects and build a security barrier. We should establish a coordinated mechanism to protect environment, development economy and improve people's livelihood. We should draw a red line of protecting ecology. In line with the principle of integrity of ecological system and positioning of functional zones, we should optimize the development of land and space, rationalize the relationship between protection and development, improve service functions of the eco-system and build an ecological security pattern with a sound structure and stable functions so as to maintain national ecological security.

Thirdly, we should improve technical support by integrating systems for monitoring, regulation and early warning. Satellite remote sensing and other technologies should be utilized to build a remote sensing monitoring system on ecology and monitor and inspect specific areas. Big data should be mined to build a national comprehensive databank for national ecological security by comprehensively utilizing such technologies as spatial analysis, information integration and internet plus, etc. Through analysis and assessment of the current situation and dynamic changes of ecological security, we can monitor and predict the future situation and the spatial distribution of ecological security, build a national monitoring and early warning system, understand the current situation and future trends of national ecological security and establish a platform for information assessment, release and response to ensure national ecological security and provide evidence for national decision making. We should formulate relevant standards to conduct overall assessment of national ecological security and release information so that the whole society can vividly understand the latest information on China's ecological environment, raise public awareness and improve local ecological security system of prewarning and protection in line with local conditions.

Fourthly, we should promote international coordination and play an important role in global ecological governance. At the global level, concerns about ecological security focus on climate change and transnational air pollution, etc., so special agencies could be set up under UN and other organizations to be responsible for formulating and implementing regulations on global ecological security within the international framework. United Nations Environment Program, the World Wildlife Fund and other international organizations could be leveraged settle international disputes over ecological security. At a regional level, there is a wide range of concerns about ecological security, including competition for water resources and mineral resources and disputes over water pollution and air pollution. Regional organizations and inter-State dialogue and consultation mechanisms could be used to coordinate ecological security issues. China now is a leading country in jointly developing the green silk road under the Belt and Road Initiative. Green economy including agi-culture, eco-industry and eco-tourism, etc. should be at the core. The Belt and Road Initiative should be taken as an opportunity to deepen the internationalization of China's environmental protection, promote international exchanges, enhance our strength in environmental capabilities, maintain ecological security and display the role of China as a big and responsible country.

第二十九章　电商进村入户念活绿色致富经

电子商务源于欧美，而盛于亚洲。自 2013 年起，我国网络零售交易额稳居世界第一，全球十大电商企业中我国占据四席，其中阿里巴巴的市场份额排名全球第一，京东商城位居第四。2016 年我国电商交易额占国民生产总值 35%，对推动供给侧结构性改革做出重要贡献。依靠"互联网+"科技创新来保护绿色生态、发展绿色生产、培养绿色生活，成为新的趋势，其中尤以农村扶贫减困、绿色发展为代表。电商将偏僻的乡村与全国乃至全球的大市场连接起来，开启了农村消贫、绿色可持续发展的新模式，是实现绿水青山"变现"金山银山的新途径，是打开农村对外窗口、启迪民智的新突破。

农村致富插上电商翅膀

我国农村电商发展如火如荼。2016 年，农村网络零售市场交易额占全国网络零售总额的 17.4%，金额高达 8945.4 亿元。以阿里巴巴旗下的淘宝为例，截至 2016 年 8 月底，1311 个"淘宝村"广泛分布在全国 18 个省份，直接创造就业机会超过 84 万个。京东集团则宣布未来五年要开 100 万家京东便利店，其中一半在农村，做到每个村都有。互联网打破行政界限，连接广域大市场，使得任何一个偏僻的地区、一个微小的个人、一种天然的禀赋都有了联通世界的能力，从而大大拓宽了脱贫致富的渠道，增添了日常生活的色彩。

电商扶贫是政策+市场的推陈出新。商务部、财政部自 2014 年起联合开展的电商进农村综合示范，是"互联网+"背景下重大的政策创新。通过市场的力量激活贫困地区的脱贫能力，"看不见的手"能更好实现政策初衷。电商进村扶贫也是电商平台的创新，自 2014 年起，阿里巴巴将农村化作为其发展的三大战略之一，京东、一亩田等互联网公司也开始把发展重点转移到农村这个主战场上来，社会上更多新资源、新手段、新资本等新生力量投入到农村、服务于农村、受益于农民。

"互联网+三农"激活经济邀民致富。电商消贫的实践在我国农村从星星之火已成燎原之势，其先行试点的示范效应尤为重要。"互联网+三农"的电商扶贫之路需要贫困地区利用本地产品资源，借助互联网工具，使产品"走出去"；另外，发掘本地特色，以民俗文化、山水风景、饮食服务等方面吸引游客，将客源"引进来"。农民自力更生帮助家乡逐步摘掉贫困帽子，脱颖成为信息时代的明星。近年来，一大批"淘宝村""淘宝镇"如雨后春笋般出现。云南省元阳县作为区位、产业均不占优势的贫困县，电商从零起步，农业产品联网成功，网上开店120余个，梯田红米销售额突破3000万元。不仅如此，山西省万荣县通过举办"万荣苹果"杯电商创新创意大赛、搭建交流平台，使电商脱贫致富的理念深入人心。

日常消费"触网"启发民智改变生活。电商打破农村销售渠道单一的局面，降低了交易成本。由于传统农村商品流通能力差、消费者购买渠道有限、缺乏维权意识，长期以来，农村消费市场假货横行、坐地起价，不少农村的小卖铺卖的都是"傍名牌"。电商的发展意味着农民选择面更广，提高辨识能力、减少开支、创造收益。同时，农村交通成本高、服务配套差，以代缴费业务为例，通信、水电、购票、挂号等均可以通过电商进行，惠农补贴、社保服务也可以通过电商渠道来实现，农村的基本生活成本及交通成本大大降低，有助于推动农业农村现代化进程。

电商进村的绿色发展隐忧

就在阿里巴巴、京东等纷纷抢滩农村电商这一蓝海市场的同时，许多问题也接踵而至。贫困地区搭上绿色发展的快车，还有许多瓶颈尚待突破。

生产规模、产品质量、市场推广的全商业线考验。生产规模方面，贫困地区以一家一户为单位的家庭作坊生产方式为主，生产规模的局限性难以满足电商模式要求的大批量产品供应的要求。

产品质量上，农产品质量参差不齐非标准化问题突出，大大影响其市场竞争力，加之贫困地区距离城市较远，物流成本较高，农民为提高收入只能降低生产成本，更加导致质量问题频发。

再到市场推广，虽然电商打破了传统商业模式下的推广壁垒，但随着竞争者的增加，如果缺乏适当的营销策略，辅以良好的服务和体验，形成品牌和口碑，农产品则很难在激烈的市场竞争中脱颖而出，"百花齐放，百家争鸣"也可能适得其反，电商扶贫与绿色发展将只能昙花一现，难以为继。

物流网络成为亟待弥补的短板。农村电商巨大的发展潜力与发展进度不匹配，资本、企业纷纷铩羽而归，其中物流是制约农村电商发展的核心。没有物流的农村，发展电商只能是空中楼阁。农村物流成本居高不下，是因为农村人口数量庞大且居住分散，订单密度较小，物流需求呈现"点多面广"的特点，但现阶段资本、技术、人力等未能达到要求；而且农资产品一般体积大、重量高、易腐烂，"距村越近，道路越窄，路况越差"的现状，导致客运货车无法通行，或者车速缓慢，"最后一公里""最初一公里"成为产品是否能够"下得去""出得来"的关键，也成为最具农村特色、追求新鲜程度的各类产品占领市场的关键。目前我国农村整体物流体系落后，缺乏专业性、现代化、大型的物流企业。物流资源不成体系，运力资源难以整合。

海量包装致使垃圾围村。电商离不开物流，物流少不了包装。农村作为一片美丽净土，呼唤发展，但更需要绿色，包装已成为制约农村电商绿色发展的瓶颈。农村电商迅速发展，快递业务量逐年增长，包装需求势头凶猛，包装垃圾问题浮现。过度包装问题尤为突出，为避免产品颠簸损坏，商家往往通过多层包装"五花大绑"以换取心理安慰、减少店铺差评；由于包装材料不环保，过度包装不仅浪费，而且带来高污染。以胶带为例，目前绝大多数快递包裹用的都是不可降解胶带，其主要成分为聚氯乙烯，需耗时百年才能降解；同时，塑料编织袋、塑料袋、纸箱等包装材料均可以多次使用，提高资源使用效率，但包装循环利用机制不健全、人们环保意识欠缺，浪费与污染形势严峻。

绿水青山"变现"的配置与引擎

政府、非政府组织、合作社等各方力量聚起来。首先，政府搭台谋发展。政府"有形的手"需在绿色发展的关键时期着力推动电商扶贫工作，加强顶层设计，明确发展目标，强化保障措施，配套优惠政策，健全组织机构；加强农村物流体系建设、乡村网点信息化改造、农村产品网络销售和人才培养，完善农村电商运营网络，引导农民树立品牌意识。其次，NGO组织聚众力。发挥NGO组织动员人力、物力、财力参与社会活动的优势，建立"线上、线下"的培训体系，帮助贫困群众开展电商创业。再次，合作社推动规范化。利用合作社制度改变小农经济的生产方式，形成规范化、标准化的运营方式，从而形成规模生产和品牌效应；通过合理的分工，以能人带动群众；集中本地与外部资源优势，促使整个区域得到自立能力与内生能力的提升。

物流、包装绿起来。在包装方面，农村电商上、下游供应链的各个环节都要避免包装材料的过度浪费，尽可能节约包装物的使用量；制定绿色包装的相关法律、法规与相关标准，鼓励创新地对现有材料择优使用，扶持绿色包装企业的研发和生产，为使用绿色包装的农户提供政策补贴，于"起点"和"终端"摒弃传统包装材料，走生态环保新道路；建立包装回收体系与制度，例如鼓励长期交易的双方通过主动完好地收集上批产品包装物来抵扣下批产品价款，形成良性循环。

在物流方面，商务部、财政部、交通部等部委加快推动农村地区交通改善，减少农村电商物流过程的时间及资金消耗；建立并完善县、乡、村三级物流配送体系，提高物流分拨及配送效率；国家通过政策、资金支持，鼓励阿里巴巴、京东等电商企业"下乡进村"开拓市场，形成先行的物流突破；建设改造县域电商公共服务中心和村级电商服务站，拉动"政、企、民"三方联动发展电商。

产品品牌亮起来。品牌效应能够提升产业及产品的溢价空间，降低生产成本，保护生态环境，实现长效发展。品牌建设在挖掘区域文脉的同时，有助于提升贫困地区的精神特质、文化气质，再造乡村文明。云南省文山市是三七原产地，产量占全国的95%，"滋补中国"品牌计划将其列为重点合作对象，带领文山这一国家级贫困县一跃成为"2016年电商消贫十佳县"榜首。贫困地区普遍存在产品丰富但品牌弱小的问题，物种资源、传统农耕文化资源、区域特色自然资源等都大为富饶，而各类产业、产品的品牌化程度极低，造成产品大面积低价出售。实施品牌消贫，利用品牌知识培训、农产品区域公用品牌战略规划设计、企业品牌创建、人才培养机制等，提高以品牌为核心的农业现代化程度，改变以生产为导向的理念与方法，从而提升企业、合作社、农户的品牌管理能力与市场竞争水平。

Chapter 29
E-commerce Enters Villages to Foster Green Development

Originating in Europe and North America, E-commerce flourishes in Asia. Starting from 2013, China's online retail transaction volume has ranked No. 1 in the world. Four of the top 10 world E-commerce companies are in China. Alibaba has the largest market share and JD.com ranks No. 4. In 2016, online traction volume accounted for 35% of China's GNP, which made important contributions to promoting structural reforms on the supply side. It has become a new trend to protect green ecology, boost green production and foster green lifestyles by relying on Internet plus and technological innovation. The case in point is rural poverty alleviation and green development. E-commerce initiates a new pattern of rural poverty reduction and green and sustainable development and blazes a new trail of turning clear water and green mountains into mountains of gold and silver by connecting remote villages with global markets, thus achieving new breakthroughs in opening up the rural area and enlightening the general public.

E-commerce Adds Wings to Rural Take-off

China's rural E-commerce has developed in full swing. In 2016, rural online retail volume reached 894.54 billion RMB and accounted for 17.4% of the total in China. Take Taobao, a subsidiary of Alibaba, as an example. By the end of August 2016, there had been 1,311 so-called Taobao Villages scattered in 18 provinces in China, which directly created 840,000 job opportunities. Jd.com announces that it will open 1 million convenience stores in the next five years with half of them located in rural China so that every village could have one store. Internets breaks down administrative barriers and connects the broader markets so that any remote region and any individual can be connected to the outside world. The channels of alleviating poverty and becoming rich has been tremendously broadened and daily rural life has been enriched.

Poverty reduction through Ecommerce is a creative measure integrating policies and market forces. The comprehensive demo project of E-commerce entering rural area jointly launched by Ministry of Commerce and Ministry of Finance in 2014 is a major policy innovation in the context of Internet plus. The market forces could be leveraged to enhance capabilities in poverty-stricken regions and the invisible hand can better fulfill the original policies intentions. E-commerce entering the village to help with poverty reduction is also an innovation by E-commerce platforms. Since 2014, Alibaba has regarded it as one of its three major strategies to occupy the rural market. JD.com, Yimutian and other internet companies began to shift their focus on the vast rural market. More social resources, measures, services and capital will be input into the countryside, which will benefit the farmers.

Using internet to resolve agriculture-related issues can help to activate rural economy and enrich the farmers. The practice of reducing poverty through E-commerce has been widely adopted in rural China on the basis of some pilot programs. Poor regions should utilize internet to sell local products. In addition, they should explore local characteristics and attract tourists with folk culture, beautiful landscapes and catering services. Farmers can become stars in the information age by helping their hometowns to shake off poverty. In recent years, a large number of towns and villages with thriving on-line business have sprung up. Yuanyang County in Yunnan Province is a poor county without any advantages in location and industrial development. However, E-commerce here has started from scratch and local agricultural produces have been sold to other parts of China through 120 online stores. The sales volume of terrace red rice has totaled 30 million RMB. Wanrong County in Shanxi Province raised public awareness of using E-commerce to shake off poverty through holding Wanrong Apple Cup Contest of Innovative & Creative Ideas on E-commerce and building exchange platforms.

Online purchase of daily necessities can enlighten the public and change their lives. E-commerce broadens the existing limited sales channels and reduces transaction cost. Commodity circulation in traditional rural areas is poor, consumers do not have many purchase options and they also lack awareness to protect their rights, so for a long time, fake products have been rampant in rural markets and commodity prices have been very high. The development of E-commerce means farmers have more options and they can distinguish genuine products from fake ones, cut costs and generate more revenues. Meanwhile, transportation cost in the countryside is very high due to poor infrastructure. Now farmers can pay their bills on telecommunication and utilities, buy tickets and make appointments with doctors online. Agricultural subsidies and social insurance services can also be provided through E-commerce. The basic living cost and transportation cost have been greatly reduced, which can promote the modernization process in rural areas.

Challenges Facing Green Development of E-commerce in Rural Areas

As Alibaba, JD.com and other e-commerce companies rush into the blue ocean of the rural markets, many problems have arisen. There are still bottlenecks to be overcome when the poverty-stricken areas are entering the fast lanes of green development.

There are challenges regarding production scale, product quality and marketing. In terms of production scale, household-based family workshop is the main form of production in poor regions. The small-scale production cannot meet the demand of E-commerce which requires the supply of large numbers of products.

The quality of agricultural produces is not uniform, which considerably affects the market competitiveness. In addition, the cost of logistics can be very high since these poor regions are far away from cities. Farmers can do nothing but reduce production cost in order to increase income, which in turn leads to frequent quality problems.

Regarding marketing, though E-commerce breaks down barriers for product promotion in the traditional business models, it is difficult for agricultural produce to stand out in the fierce market competition if there is no proper marketing strategy, excellent service, experience, brand or reputation. Poverty reduction through E-commerce and green development can only be a flash in the pan and cannot be sustained.

Logistics network has become a drawback which should be overcome urgently. The speed of E-commerce development cannot match the tremendous potential and the capital and E-commerce companies suffered frustration and setback. Logistics is the core factor restricting the development of rural E-commerce. If there is no adequate logistics in rural China, developing E-commerce can only be a castle in the air. The high cost of rural logistics is due to large butwidely scattered rural population and small density of orders, thus causing high pressure to logistics. At the current stage, capital, technology and human resources cannot meet the requirement. In addition, agricultural produces are mainly bulky, heavy and perishable. The nearer to the village, the narrower the road and the worse the road conditions, so the trucks cannot get through or have to slow down. The last mile and the first mile become the key to whether the agricultural produce can be transported to the outside or the outside products can be delivered to rural consumers. In particular, fresh agricultural produce with local color bear the brunt. At present, the logistics system is very backward in rural China and there are no professional, modern and large logistics companies. It is difficult to integrate fragmented logistics resources.

Massive packaging has caused a large amount of garbage in villages. Packaging is a must for logistics. Beautiful countryside called for green development. However, packaging has become a bottleneck restricting the green development of rural E-commerce. The rapid development of rural E-commerce leads to huge increase of express delivery services which generate a great demand for packaging. Excessive packaging has become a serious issue. In order to prevent their products from being damaged during delivery process and receive favorable comments from the buyers, the sellers often package their products excessively for psychological comfort. Because the packaging materials are not environmental-friendly, excessive packaging results in not only waste, but also high pollution. Take adhesive tapes as an example. Currently, most of the delivery parcels contain non-degradable tapes whose main content is PVC. It takes over 100 years for the material to degrade. Meanwhile, plastic bags, carton boxes and other materials can be used repeatedly to increase efficiency of resource utilization. However, the recycling mechanism of packages is not well-established and people lack environmental awareness, causing serious waste and pollution.

Ways of "Monetizing" Clear Waters and Green Mountains

Governments at all levels should join hands with NGOs, cooperatives and other social forces. Firstly, government departments should build a platform for cooperation and development. "The visible hand" must spare no efforts to promote poverty reduction through E-commerce at the crucial stage of green development by enhancing top-level design, clarifying development goals, providing preferential policies and supporting measures and strengthening organizational arrangement. The government should build rural logistics systems, improve information network at village level, enhance rural products marketing, train talents, better rural E-commerce operation network and guide farmers to raise brand awareness. Secondly, NGOs should play an important role in mobilizing human, material and financial resources, setting up online and offline training systems and helping poor farmers to start E-businesses. Thirdly, cooperatives can transform the production mode of small-scale farming and develop standardized operational mode so as to establish large-scale production and brand effect. Capable farmers can lead their fellow villages through reasonable division of labor. Local and external resources should be consolidated to enhance local capacity building.

Logistics and packaging should be green. Regarding packaging, all the links in downstream and upstream supply chain in rural E-commerce should avoid waste caused by packaging materials by reducing their use. The government should formulate laws, regulations and standards on green packaging, encourage the use of high-quality materials in an innovative way, support the R&D of green packaging companies, provide policy subsidies to farmers who use green packaging and advocate the development approach of ecology and environmental protection by discarding traditional packaging materials at the starting points and at the ending points. Packaging material recycling systems and mechanisms should be established. For example, packaging materials can be used as cash for the next batch of products to be purchased to generate a virtuous cycle.

Regarding logistics, ministries of commerce, finance and transportation, etc. should promote the improvement of road conditions in rural areas, reduce delivery time and traffic barriers, build and improve a three-tier logistics system at county, township and village levels and increase logistics efficiency. The central government has encouraged Alibaba, JD.com and other E-commerce companies to tap rural markets with policy and financial support so as to achieve logistical breakthrough. County-level E-commerce public service centers and village-level service stations should be established. The government should work together with the private sector to develop E-commerce.

Brand awareness should be raised. Brand effect can help to reduce production cost, increase product value, protect ecology and environment and achieve long-term development. Brand building can fully take advantage of local culture and help to improve the spiritual and cultural image of poverty-stricken regions so as to rebuild rural civilization. Wenshan City in Yunnan Province is the place of origin for notoginseng and boasts 95% of the national output. The Nutritious China Brand Initiative selected notoginseng as a key target for cooperation and helps Wenshan City, a national level poor county, to become one of the top 10 counties which shook off poverty through E-commerce in 2016. There is a rich variety of products, species resources, traditional farming cultural resources and natural resources with regional features, etc.in poor regions, yet their brands are not well-known. Therefore, the products are sold at a very low price. Brand awareness should be raised to reduce poverty through brands. Local talents should be trained on brand knowledge, brand planning and design, corporate brand creation, etc. so as to improve level of agricultural modernization with brands at the core, change the production-oriented concept and method and improve companies, cooperatives and farmers' brand management capability and market competitiveness of products.

第三十章　环保督察整改倒逼绿色转型

自 2016 年年初中央环保督察组在河北开展督察试点以来，2016～2017 年间，中央环保督察组先后分四批入驻全国 31 个省（区市），对全国环境执法监管进行督察。开展环保督察，是党中央、国务院推进生态文明建设的重要抓手，为落实地方政府环境保护主体责任，优化环保体制，促进经济健康发展提供了有效的制度保障。

回看督察过程，成果显著。很多老百姓对督察组十分期待，有的直言："督察组早两年来就好了。"但同时，社会上也有一些负面情绪，甚至出现"环保一刀切，百姓生计堪忧""要环保不要发展"等质疑。如何消除民众质疑，是正确认识环保督察及其成果，保障督察延续、协调环保与经济发展、改善民生关系的关键。

环保督企督政激浊扬清

手持"尚方宝剑"的中央环保督察组，其目标在于重点盯住中央高度关注、群众反映强烈、社会影响恶劣的突出环境问题及其处理情况；重点检查环境质量呈现恶化趋势的区域流域及其整治情况；重点督办人民群众反映的身边环境问题的立行立改情况；重点督察地方党委和政府及其有关部门环保不作为、乱作为情况；重点推动地方落实环境保护党政同责、一岗双责、严肃问责等工作情况。所到之处，解决了百姓关注的突出环境问题，有力地震慑了污染企业，有效地鞭策了地方政府。

两年督察全覆盖，问责处罚不手软。四批环保督察基本上实现了除港澳台地区之外的 31 个省（区市）的全覆盖。中央环保督察接受群众来电来信举报之多，约谈、问责人数之多，处罚力度之大前所未有，反映出政府对督察的重视程度。铁腕之下，污染企业有的被罚款，有的被停业整顿，有的被直接取缔，很多没有查到的企业在威慑之下主动整改，环保督察之严厉有效威慑了违法违规企业。

从"督企"到"督政"，实现党政同责。长期以来，环境监管只注重"督企"而忽视"督政"，污染型企业受到地方政府保护的问题较为突出。因此，中央环保督察组在多地掀起一场治污问责的风暴，明确指出各级地方党委和政府存在的问题，指明其环境改善的责任，推动落实环境保护"党政同责""一岗双责"。具体来说，主要督察以下内容：各级政府贯彻落实党中央、国务院环境保护重大决策部署及落实国家环保相关法律法规、行动计划等文件的落实以及解决和处理突出环境问题、改善环境质量等方面的情况。督察结束后，重点问题报告中央，督察结果移交组织部，结果将作为被督察地方领导班子和领导干部考核评价任免的重要依据。整个流程反映出"严格依法办事""严格政策落实""督察回头评估检查"的新督察面貌。环保督察工作常态化有效推动了以改善环境质量为核心的环境管理转型，加强了各级政府环保体系建设，为监督地方政府严格执行环境法律法规和政策，建设长期有效的环保监督体系打下了坚实的基础。

"大手笔"引来诸多争议

全国范围内，四批中央环保督察均已结束，严格的环保标准之下，众多企业停产、工人停工，家庭生计堪忧。"督察覆盖面过大""环保督察影响地方经济"的说法甚嚣尘上。

环保督察铁腕手段引来发展"阵痛"。环保督查的严厉查处过程中存在的企业关停、百姓自家作坊查处等情况造成了或轻或重的就业与民生问题，甚至影响了地方经济发展，出现增速减缓乃至明显的下滑现象。中央展开督察的初衷在于提高基层环境治理能力，改善环境质量，回应人民群众对环境问题的关切。在督察过程中坚持强硬态度，对违法排污企业坚决取缔，短期内不可避免地会对经济发展造成影响，"阵痛"在所难免。

然而，不遇高山不必开路，不见险滩无以架桥，越是困难的转型期，越是历史的机遇期。严格的环保督察措施，加快了落后产能淘汰，促进产业结构转型，推动供给侧改革，达到经济发展与环境保护双赢的目的。另外，从行业规范性和社会经济良性发展的角度来看，"散污乱"的企业由于生产成本低，长期扰乱行业售价、扰乱用工价格和标准、扰乱合法纳税环境，使得劣币驱逐良币，行业无法持续健康发展。查处关停高污染、高耗能的企业是对过去执法不严的纠正，污染企业利益受损只是"一时之痛"。长远来看，铁腕治理环境问题是为未来发展积蓄力量。

层层加码"一刀切"导致地方质疑不断。自环保督察之风兴起，企业"关停潮"随之而来。一些地方出于最简单、最省事的办法考虑，以产业划线、以区域设界，不管是否合法，督察一来全部关停，实行环保管控"一刀切"。这种简单粗暴的方法，看似见效快，对于上级督察有所交代，但对于推进环保与经济协同发展却大为不利。"一刀切"式的执法，换来的只能是民怨沸腾。事实上，环保整改、环保督察，目的都在于改善环境，维护广大人民群众的利益，既是政治任务，也是民心工程。环保督察绝不提倡"一刀切"的处理方式，不允许为了敷衍环保督察而一味求快。对待不同的污染企业要分类处理，对于有严重污染且又不符合相关法律法规的企业，坚决打击，对与群众密切相关的、污染排放较小且通过短时间整改能达到环保标准的相关行业企业、设施场所，不宜"一刀切"关停，相反，应着力加强整改整治，使之达到环境标准，使之尽快恢复经营。

环保督察常态化　绿水青山共为邻

引领常态化，推进法治化。中央环保督察打响了我国污染督察的响亮一枪，"先污染后治理"已经过去，"懂环保再生产"的时代已经来临。督察效果的保持要靠环保督察的长效化和常态化。当中央环保督察组结束这几批的全国环保督察行动，环保督察的重任就要落到地方身上。各省级政府应效仿中央环保督察组，组建省级环保督察组，并下沉到各个市县乡村，严查污染。

徒法不足以自行，良法还需善治。从推进生态治理体系和治理能力现代化的角度看，目前的环保督察制度还存在缺陷与法治困境，需要运用法治思维与法治方式将环保督察纳入法治轨道。环保法的威慑作用也需要环保督察的持续发力。因此，一方面要完善环保督察的法律依据、提升环保督察专门机构的法律地位、规范环保督察问责程序；另一方面要求地方政府提升监管和执法水平，必须对违法企业零容忍，依法严厉打击；同时，还要善于总结经验，将督察程序和规范上升到法律法规层面，确保环保督察的法治化运作。

警惕不作为，叫停乱作为。即使在严格的环保整治要求和严厉的环保问责压力下，仍有一些地方政府存在环保不作为、乱作为的现象。针对污染企业，只是罚款了事，不从根本上整改，实际上是纵容了污染行为。还有些地方，看上去采取了很多治理措施，但都流于表面形式，治本工程严重滞后。更有甚者，为了应付督察弄虚作假，提供虚假数据。针对这些行为，应该展开专项督察，坚决抵制地方政府的不作为、乱作为现象，做到有法必依、执法必严、违法必究。既要强调执法的严肃性，又要反对不加分析"一刀切"，将发展与环保片面对立的做法，警惕看似铁面实则懒政的行为。

民意要听取，民生要关怀。受到环保督察风暴冲击的个体经营者，由于利益受损，容易受到煽动，所以在环保督察过程中督察组一定要善于倾听群众的声音，要多听难处、多做普法、多给关怀；警惕基层民怨，高度关注"丢了饭碗""遭受暴力执法"等基层民生民权问题。对环保信访件立查立改，坚持疏堵结合，强化分类处置，对于一些民生老店，允许边营业边整改。法不外乎人情，严格执法并不意味着横眉冷对，也不意味着政府要站在人民的对立面，而是我们在追求和谐、保留人性化的基础上坚守原则。环保督察要想达到理想效果，离不开人民群众的参与和支持，始终不能偏离为人民服务的宗旨，环保督察任重而道远。

Chapter 30
Environmental Inspection Drives Green Transformation

Since environmental inspection teams from the Central Government launched pilot inspection programs in early 2016, four rounds of similar inspections have been carried out in 31 provinces, municipalities under central administration and autonomous regions to supervise national environmental law enforcement. Conducting environmental inspection is an important leverage for the CPC Central Committee and the State Council to promote ecological progress, which can provide institutional guarantee for holding local governments accountable for environmental protection, optimizing environmental protection system and promoting healthy economic development.

Great achievement has been made in the inspection process. The general public sings high praises for the inspection teams. Some commented frankly, "The inspection teams should have been dispatched two years before". However, some negative opinions have arisen. Even there are such doubts that "environmental protection endangers people's livelihood" and that "environmental protection is carried out at the expense of development".

Environmental Protection Inspections Motivating Enterprises and Government Departments to Be More Environmental-friendly

Entrusted by the Central Government, the inspection teams aim to focus on those notorious environmental problems causing strong public responses and bad social impact which the Central Government also attaches great importance to. They mainly inspect those regions with deteriorating environmental quality and local improvement measures, urge local governments to resolve environmental issues reported by the general public, supervise environmental inaction or illegal behaviors of local party committees and governments and promote the establishment of an accountability system for local officials. Wherever they go, they have solved serious environmental problems, deterred polluting companies and effectively pressured local governments to take action.

Two years of comprehensive inspections have resulted in severe punishment for those wrongdoers. Four rounds of inspections have basically covered 31 provinces, municipalities and autonomous regions in mainland China. The central inspection teams have received many whistle-blowing letters and emails and questioned and punished many officials at an unprecedented level, which reflects the central government's attention on inspections. Some polluting enterprises have been fined, suspended or even closed directly. Those companies who have not been inspected have actively made improvement. The strict environmental inspections have effectively deterred those enterprises which violate laws and regulations.

Not only enterprises but also party committees and governments at all levels have been inspected and held accountable. For a long time, environmental inspections have just focused on enterprises and neglected governments and polluting enterprises have been protected by local governments. Therefore, the central inspection teams have launched an accountability storm in different regions, identified the existing problems of local party committees and governments, clarified their responsibilities to improve environment and implement the mechanism of equal responsibilities of party committees and local governments and "one post with two responsibilities". To be specific, the inspection teams mainly inspect the following aspects: whether governments at all levels have implemented the major decisions of the central government on environmental protection, what measures they have taken to implement relevant laws, regulations, action plans and other documents, how they have solved severe environmental problems and improved environmental quality, etc. At the end of the inspections, important issues will be reported to the central government and inspection results will be reported to organizational departments and used as an important criterion for assessing and promoting officials. The whole inspection process reflects law-abiding inspections to check the strict implementation of policies and continuous assessment after the inspections. The normalization of environmental inspections has effectively promoted the transformation of environmental management with environmental quality at the core, strengthened the construction of environmental protection systems at all levels and laid a solid foundation in establishing a long-term and effective environmental inspection system to supervise local governments in strict implementation of environmental laws, regulations and policies.

Drastic Measures Trigger Controversies

The fourth round of central environmental nationwide inspection has come to an end. In line with strict environmental standards, many companies have been closed and workers have become idle, which has affected their livelihood. There have been rampant opposing opinions which hold "the wide-ranging environmental inspection has affected local economy".

Stringent measures taken during the inspections have caused painful consequences. Some companies and family workshops have been closed, triggering unemployment and even adverse impact on local economic development. Economic growth has slowed down or even declined obviously. The purpose of the inspections is to enhance environmental governance capabilities at the grassroot level, improve environmental quality and respond to people's concerns over environmental problems. Tough measures have been taken to close those polluting enterprises. It is inevitable that these measures will undoubtedly affect local economic development in the short term.

However, every cloud has a silver lining. The difficult period of transformation is also full of historic opportunities. Tough measures taken during the inspections have sped up the elimination of backward production capacities, promoted transformation of industrial structures, pushed forward the reform on the supply side and achieved win-win in economic development and environmental protection. In addition, these small and disorderly polluting companies have long disrupted product prices and labor salaries and standards and evaded taxes because their production cost is very low. Due to unfair competition, Chinese industries cannot develop in a healthy way. Punishing these highly polluting enterprises with high energy consumption is a rectification of weak law enforcement in the past. The loss these polluting companies suffer is only temporary. In the long run, tacking environmental pollution is to accumulate strength for future development.

Drastic measures to tackle pollution have caused many controversies. Since the beginning of environmental inspection, many enterprises have been shut down. When inspection teams come, some local governments adopt the simplest method of closing enterprises in certain industries and regions no matter those companies are legal or not in order to meet environmental requirement. This simple and crude method looks very effective, but it is harmful to the coordinated development of environmental protection and economy. This kind of rigid law enforcement has trigged widespread resentment. In fact, the purpose of environmental inspection is to improve environment. Safeguarding people's interests is not only a political task, but also a project to benefit people. The simple and crude method of imposing uniformity in all cases by closing all factories must be prohibited. Different measures should be taken regarding different polluting companies. Those companies causing heavy pollution against relevant laws and regulations should be closed resolutely. Those companies or facilities with light pollution should not be simply closed if they can meet environmental standards after retrofit.

Conducting Environmental Inspection Regularly to Preserve Clear Water and Green Mountains

Environmental inspected should be normalized based on the rule of law. The central environmental inspection has fired a loud shot. The era of "treatment after pollution" has become a thing of the past and the era of "environmental protection before production" has come. We should establish a long-term and regular mechanism of environmental inspection in order to ensure the effect. When the four rounds of central inspections come to an end, the important task of environmental inspection will be performed by local governments. Provincial-level inspection teams should be set up and sent to cities, counties, townships and villages to conduct regular inspections and prevent pollution.

Laws alone cannot produce good results and good laws require good governance. From the perspective of promoting modernization of ecological governance system and governance capabilities, there exist drawbacks and legal dilemma in the current environmental inspection system. Environmental inspection should be incorporated into the system of the rule of law. The deterrent effect of Law on Environmental Protection can be enhanced by environmental inspection. Therefore, on one hand, we need to improve the legal basis of environmental inspection, enhance the legal status of environmental inspection agencies and standardize the accountability procedures. On the other hand, local governments should improve their supervision and law enforcement capabilities, adopt a zero-tolerance attitude towards enterprises violating the law and punish them severely according to law. Meanwhile, we should summarize lessons and experiences and incorporate inspection procedures and standards into laws and regulations to ensure the operation of environmental inspection in line with the principle of the rule of law.

We must guard against government inaction and prohibit arbitrary actions by governments. Even under the pressure of environmental governance and accountability, some local governments still take no measures or take arbitrary measures. They only ask the polluting enterprises to pay fines and do not take measures for fundamental changes, which actually pampers polluters. In some regions, it seems that the governments have done a lot, but these measures are only superficial. Even worse, some local governments falsify data to cheat the inspectors. Special inspections should be carried out in response to these behaviors. Local governments should adhere to the principle that "there must be laws to go by, the laws must be observed and strictly enforced and law-breakers must be prosecuted". They should enforce the law in a serious manner and coordinate the relationship between economic development and environmental protection. The simple and crude method of closing all companies before inspections seems very resolute and iron-handed, but it actually means laziness.

We must listen to public opinions and care about people's livelihood. Those self-employed businesses are easily incited because their interests are affected in environmental inspections. Therefore, in the inspection process, inspectors must listen to their voices, understand their difficulties, care about their interests and popularize relevant laws and regulations. We must guard against public resentment at the grassroot level and pay due attention to people's livelihood, rights and interests. People's complaints must be handled immediately and wrongdoings should be corrected. Different situations should be treated differently. Some time-honored brands should be allowed to do business while adopting some rectifying measures even if they fail to meet environmental standards. Laws should be enforced strictly, but we should also be considerate. Strict law enforcement does not mean we must be hostile to the polluters. Nor does it mean that governments are on the opposing side of the people. Rather we should adhere to the principle and enforce the law in a harmonious and humane way. Without people's engagement and support, environmental inspection cannot be conducted successfully. The principle of serving the people should be abided by at any time. Environmental inspection still has a long way to go.